Unmaking the Bomb

Unmaking the Bomb

A Fissile Material Approach to Nuclear
Disarmament and Nonproliferation

Harold A. Feiveson, Alexander Glaser, Zia Mian, and
Frank N. von Hippel

The MIT Press
Cambridge, Massachusetts
London, England

First MIT Press paperback edition, 2016

© 2014 Massachusetts Institute of Technology

This book was set in Sabon LT Std 10/13pt by Toppan Best-set Premedia Limited, Hong Kong.

Library of Congress Cataloging-in-Publication Data.

Feiveson, Harold A.
Unmaking the bomb : a fissile material approach to nuclear disarmament and nonproliferation / Harold Feiveson, Alexander Glaser, Zia Mian, and Frank N. von Hippel.
  pages   cm
Includes bibliographical references and index.
ISBN 978-0-262-02774-8 (hardcover: alk. paper)
ISBN 978-0-262-52972-3 (paperback),   1. Nuclear nonproliferation.
2. Nuclear disarmament.   3. Nuclear fuels—Management.   4. Nuclear fuels—Security measures.   I. Glaser, Alexander, 1969–   II. Mian, Zia.   III. Von Hippel, Frank.   IV. Title.
JZ5675.F45 2014
327.1'747—dc23
                              2014003654

Si vis pacem para pacem—if you want peace prepare for peace.

—Joseph Rotblat (1908–2005), Manhattan Project physicist and 1995 Nobel Peace Prize Laureate for his "efforts to diminish the part played by nuclear arms in international politics and, in the longer run, to eliminate such arms"

The day that the last nuclear weapon on earth was destroyed would be a great day. It would be a day for celebrations. We would have given substance to our choice to create the human future.

—Jonathan Schell (1943–2014), author

Contents

Preface

This book builds on many years of research, writing, teaching, and policy activism that have aimed to highlight and confront the challenges to our civilization posed by the existence, continued production, and use of fissile materials—primarily, plutonium and highly enriched uranium—the essential ingredients of nuclear weapons.

Fissile materials did not exist before 1940, even in the laboratory. The creation of the current global stockpiles of these materials over the past seven decades was a result of decisions by a small number of states to rely on nuclear weapons and on specific technology choices made for nuclear power. As of 2013, there was enough fissile material for a total of over 100,000 nuclear weapons. Almost all of this material was held by the nine states with nuclear weapons, with a very small fraction held under international monitoring by a few non-weapon states.

Many people and many policy makers have come to assume that the world can continue indefinitely with thousands of nuclear weapons and huge quantities of weapon-usable materials being produced, stored, and used at many locations around the world without disaster occurring. It defies the odds, however, to believe that, because nuclear weapons were not used during the Cold War and because there have been no terrorist nuclear explosions, nothing terrible will ever happen.

The book describes the history, production, current stockpiles, and use of fissile materials in nuclear weapons and in nuclear reactor fuel and proposes a set of policies to drastically reduce the quantities and the number of locations around the world where these materials can be found with a view to their total elimination. We hope that it will serve to make more credible to governments and to citizens the need for and opportunities that exist for progress on this policy agenda.

The goal that we pursue is not new. Even before the first nuclear weapon was built, some scientists warned of the grave new dangers posed by nuclear materials. One such voice was Niels Bohr, a pioneer of modern physics who at the turn of the twentieth century discovered the basic structure of the atom and in the 1930s explained the mechanism of nuclear fission, the process that unleashes the enormous energy that powers both nuclear weapons and nuclear reactors. In a July 1944 letter to U.S. President Franklin D. Roosevelt, Bohr cautioned that "a weapon of an unparalleled power is being created which will completely change all future conditions of warfare. Unless, indeed, some agreement about the control of the use of the new active materials can be obtained in due time, any temporary advantage, however great, may be outweighed by a perpetual menace to human security."

Efforts aimed at addressing the threat were stalled by the animosities of the Cold War. In the two decades since this world-threatening conflict ended, the determined efforts of nuclear weapon and nuclear energy establishments and their supporters to sustain the nuclear enterprise have combined with a lack of public focus on these issues to allow only fitful progress.

The authors, all physicists, are members of Princeton University's Program on Science and Global Security and have worked separately and together on fissile material issues for many years. Since 2006, the authors have been involved in organizing and leading the International Panel on Fissile Materials (IPFM) through which independent experts, scholars, activists, and former diplomats from eighteen countries have been educating their governments and the international community about the urgent need and opportunities to reduce the dangers from these materials.

This book draws heavily on studies and reports by the IPFM and the path-breaking work by David Albright, Frans Berkhout, and William Walker in the 1990s. Our understanding of these issues has been strengthened and deepened by discussions and debates with our IPFM colleagues and with many other analysts and activists around the world who share our dedication to the advancement of policies that will ensure fissile materials will not be, as Niels Bohr feared, a "perpetual menace" to humankind. It is to them that we dedicate this book.

Harold Feiveson, Alexander Glaser, Zia Mian, Frank von Hippel
Princeton, 2014

Figure I.1

Map of enrichment and reprocessing facilities worldwide

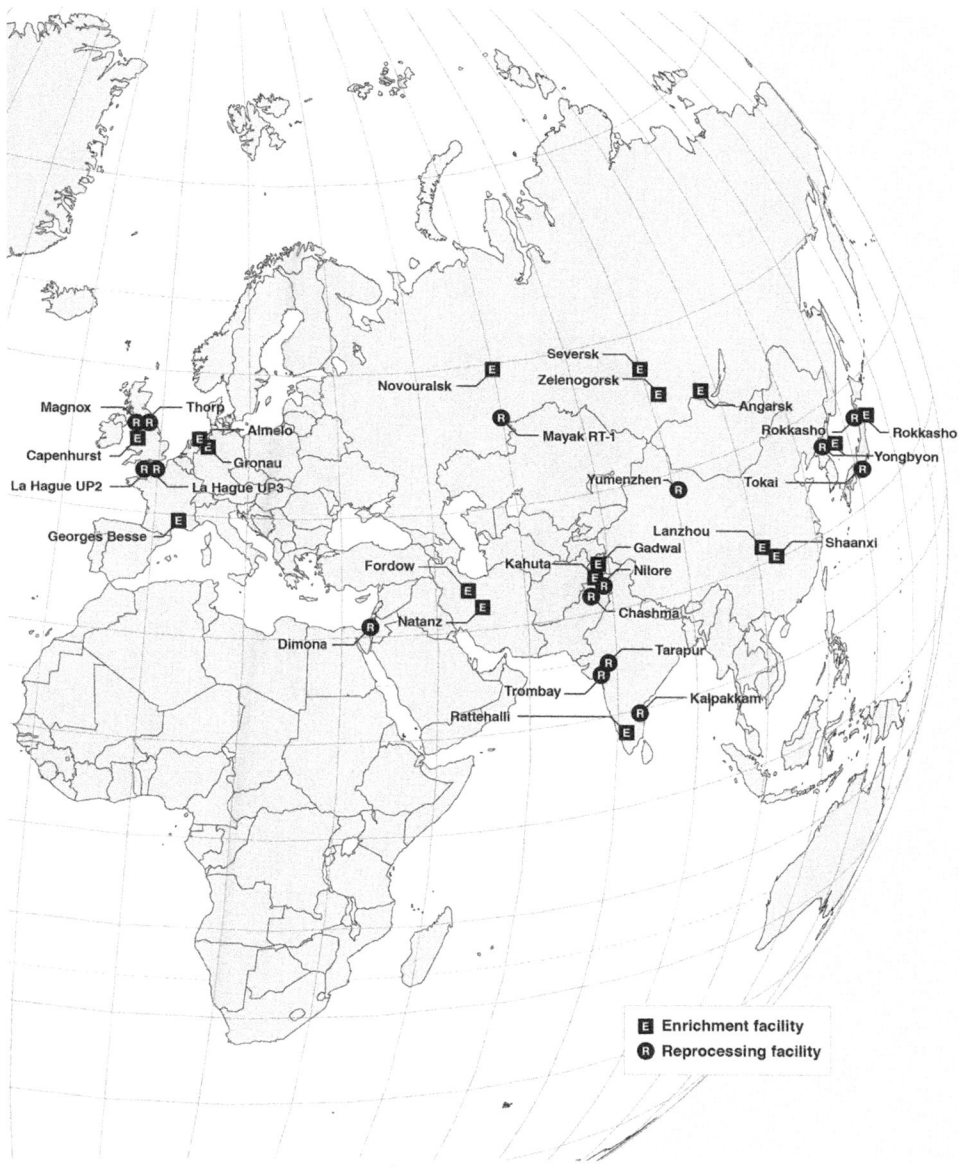

Figure I.1

(continued)

1

Introduction

With the invention and use of the atomic bomb, humanity developed the means to end civilization within hours. While nuclear weapons have not been used in war since then, threats and preparations to use them and the pursuit of the capability to produce them have cast a terrible shadow over the world. It is hard to see how disaster can be forever forestalled in a world that now has nine states holding nuclear weapons, with additional countries and terrorist groups possibly seeking to acquire them.

Earlier historical epochs were called the Stone Age and the Iron Age. For almost seven decades, the world has lived in a Fissile Material Age. Fissile materials can sustain an explosive nuclear fission chain reaction that releases enormous energy in the form of blast, heat, and radiation. The energy from such explosions also is used to ignite the nuclear fusion reactions in modern thermonuclear weapons (hydrogen bombs). All nuclear warheads contain at least a few kilograms of either plutonium or highly enriched uranium (HEU) and often both. If we are to reduce the threat from nuclear weapons, we must deal with the dangers posed by the production, stockpiling, and use of fissile materials.

The first nuclear explosion—the Trinity test carried out by the United States in the Alamogordo Desert in southern New Mexico on July 16, 1945—was a test of the plutonium bomb that was exploded twenty-four days later over Nagasaki. A bomb made from HEU, based on a different design, so simple that it did not require an explosive test, was used against Hiroshima on August 6, 1945. In each of these two bombs about one kilogram of fissile material fissioned, releasing in about one microsecond energy equivalent to the explosion of nearly 18,000 tons of high explosives.

When he learned of the Hiroshima bombing, Harry Truman, the president of the United States, described it as "the greatest thing in history."[1]

In his public announcement, he described the weapon as "a new and revolutionary increase in destruction," with "more than two thousand times the blast power of the . . . largest bomb ever yet used in the history of warfare."[2] He identified Oak Ridge (Tennessee) and Richland (Washington) as the sites where the United States had "been making materials to be used in producing the greatest destructive force in history."

Truman also made public the enormous scale, secrecy, and cost of the effort of making the fissile materials for the bombs:

We now have two great plants and many lesser works devoted to the production of atomic power. Employment during peak construction numbered 125,000 and over 65,000 individuals are even now engaged in operating the plants. Many have worked there for two and a half years. Few know what they have been producing. They see great quantities of material going in and they see nothing coming out of these plants, for the physical size of the explosive charge is exceedingly small. We have spent two billion dollars on the greatest scientific gamble in history—we won.[3]

The huge secret effort described by President Truman, code-named the Manhattan Project, was the first successful production of fissile materials in sufficient quantity for a nuclear weapon. Seven decades later, that remains the most significant challenge to acquiring nuclear weapons.

From the outset of the fissile material age prominent scientists had warned of the grave dangers that these new materials and nuclear weapons posed to the world (figure 1.1). In January 1946, in its first resolution, the General Assembly of the newly formed United Nations called for plans for the elimination of nuclear weapons and for control of atomic energy to ensure it was used only for peaceful purposes.[4] Competing plans were developed by the United States and Soviet Union but Cold War suspicions prevented agreement.

Since the failure of the first efforts to ban nuclear weapons and control the use of fissile materials, nine other states have successfully followed the United States and produced nuclear weapons through the uranium enrichment or plutonium separation routes to the bomb. In most cases, weapon states have pursued production of both HEU and plutonium.

The physical and chemical technologies that allow uranium enrichment and plutonium separation also have been mastered by many other states that have considered but decided not to build nuclear weapons. These technologies are now within the reach of a growing number of states with modest scientific and industrial capacity. As fissile material

Figure 1.1

Seven decades ago, scientists began warning of the dangers of fissile materials and their use in nuclear weapons. From left: Niels Bohr in 1944 cautioned that fissile materials could become a "perpetual menace" to humankind;[a] James Franck in 1945 led a group of Manhattan Project scientists in arguing that "the development of nuclear power is fraught with infinitely greater dangers than were all the inventions of the past";[b] Albert Einstein in 1955 joined philosopher Bertrand Russell and others to issue the Russell-Einstein manifesto calling for the abolition of nuclear weapons and launching the Pugwash movement of scientists for nuclear disarmament;[c] Isidor Rabi in 1949 advised the United States government not to pursue thermonuclear weapons since "The fact that no limits exist to the destructiveness of this weapon makes its very existence and the knowledge of its construction a danger to humanity as a whole."[d]

Photo: Princeton, New Jersey, October 3, 1954.

Source: Metropolitan Photo Service photographer. From the Shelby White and Leon Levy Archives Center, Institute for Advanced Study, Princeton, NJ, USA.

Notes: [a]Niels Bohr, "Memorandum to President Roosevelt," July 3, 1944, in Niels Bohr, *Collected Works, Volume 11: The Political Arena (1934–1961)*, ed. Finn Aaserud (Amsterdam: Elsevier, 2005), 101–108; [b]James Franck et al., Report of the Committee on Political and Social Problems Manhattan Project (The Franck Report) (University of Chicago, June 11, 1945); [c]Russell-Einstein manifesto, July 9, 1955; [d]Enrico Fermi and I.I. Rabi, "An Opinion on the Development of the Super," in United States Atomic Energy Commission General Advisory Committee, Report on the "Super," October 30, 1949, reprinted in Williams and Cantelon, *The American Atom*, 120–127.

production technologies spread, so do the opportunities for the spread of nuclear weapons to more countries.

Fissile material production still appears an unlikely route to the bomb, however, for nonstate groups that may be motivated to carry out nuclear terrorist attacks. Such groups would more plausibly try to acquire an existing weapon or stocks of fissile material from which they could fabricate a nuclear explosive—a powerful additional reason for states to seek the elimination of nuclear weapons and fissile material stocks.

At the end of World War II, the United States was the only country with nuclear weapons, and the amount of highly enriched uranium and plutonium that it had produced was on the order of a hundred kilograms in total—enough for only a few nuclear explosives. During the Cold War, the number of nuclear weapons grew to the tens of thousands and the size of stockpiles of fissile materials grew to millions of kilograms (thousands of tons—throughout this book, tons refer to metric tons, with one metric ton corresponding to 1,000 kg or about 2,205 pounds). Globally, the number of nuclear warheads peaked in the late 1980s at over 65,000. To make possible such large numbers of nuclear warheads, the weapon states collectively produced for weapon purposes over 2,000 tons of HEU and about 250 tons of separated plutonium.

With the end of the Cold War, the number of nuclear weapons has declined and is likely to fall further. Even so, in 2013 there were still about 17,000 nuclear warheads worldwide, with almost half of them in line to be dismantled over the next few decades. More than 90 percent of these warheads are held by the United States and Russia. The United Kingdom, France, China, Israel, India, Pakistan, and North Korea have about 1,000 operational warheads between them and perhaps several hundred that have been withdrawn from service. South Africa dismantled its nuclear weapons in the 1990s and joined the Non-Proliferation Treaty as a nonnuclear weapon state.

The dismantlement of tens of thousands of Cold War era warheads has left the weapon states with large stockpiles of excess fissile material, which, if not eliminated, could be used for weapons again. Making nuclear weapons reductions irreversible requires disposing of this fissile material and ending the production of fissile material for weapons in those few countries where it is still going on.

The problem is larger than this, however. Some countries use (or plan to use) plutonium in civilian nuclear power reactors to make electricity, and use highly enriched uranium in civilian and military nuclear research reactors and for military naval propulsion. The largest stockpile of fissile material associated with such activities is the civilian plutonium separated from power reactor spent fuel. All of this material is weapon-usable. To reliably assure against its use for weapons will require an end to the production and use of fissile materials for reactor fuels and the disposal of existing stocks currently earmarked for such fuels.

The goal of this book is to lay out the technical basis for practical policy initiatives that would step-by-step cap, reduce, and eventually eliminate the global stockpile of about 1,900 tons of weapon-usable fissile material including material in weapons or recovered from dismantled weapons, the plutonium used in civilian nuclear power programs, and the HEU in military and civilian research and naval reactor fuel stockpiles. Such initiatives are critical to support deep reductions and ultimate elimination of all nuclear warheads, to make disarmament more difficult to reverse, to raise the barriers to nuclear weapon proliferation to countries that do not have them, and to prevent possible nuclear-weapon acquisition by terrorist groups. They complement the traditional nuclear arms control and nonproliferation agenda, which has focused on capping and reducing deployed nuclear warheads and delivery systems, limiting the testing of such weapons, and international monitoring of civilian nuclear energy programs to deter and detect their potential use for weapons.

How the Nuclear World Emerged

Part I of this book, comprising the next four chapters, provides essential background for understanding the fissile material problem. We summarize this background briefly below.

In chapter 2, we describe the processes for producing fissile materials, how these materials are used in nuclear weapons, and the current national and global stockpiles.

The splitting or "fission" of heavy uranium nuclei was discovered just before World War II, in December 1938, in Germany. Not long afterward, in March 1940, two refugee physicists, Otto Frisch and Rudolf Peierls at the University of Birmingham, England, wrote a technical memorandum alerting the British government that an explosive nuclear fission chain reaction might be possible in a small mass of nearly pure uranium-235.[5] (Natural uranium consists of 0.7 percent uranium-235 and 99.3 percent uranium-238.) They also noted that "effective methods for the separation of isotopes have been developed recently" that could allow uranium-235 separation from natural uranium on a sufficiently large scale to permit construction of an atomic bomb. When it came to the attention of U.S. scientists and policy makers, this memo galvanized the U.S. bomb program.

While the U.S. effort to design the atomic bomb, led by J. Robert Oppenheimer at Los Alamos, has captured most attention from historians, the largest investment of resources and people in the Manhattan Project was the effort in Oak Ridge, Tennessee, to separate the minor chain-reacting isotope uranium-235 from natural uranium. Three different techniques were developed, but the one that was adopted at the end of the war was gaseous diffusion. This involves the passage of gaseous uranium hexafluoride through thousands of porous barriers with the stream becoming slightly more enriched in uranium-235 at each stage, because the molecules carrying the lighter uranium-235 atoms pass through the barriers slightly more quickly. Gaseous diffusion facilities became enormous and consumed vast amounts of power.

Modern enrichment plants rely on a much more efficient isotope separation technology, gas centrifuges. A gas of uranium hexafluoride is spun at high speed inside a long vertical cylinder so that the molecules carrying the heavier uranium-238 atoms are pressed more tightly against the wall; combined with an axial circulation of the gas in the centrifuge rotor, this process can be used to extract two streams of uranium from the machine, one slightly enriched and one depleted in uranium-235. By connecting many such centrifuges in series and parallel, uranium can be enriched to any desired level, including "weapon-grade," which contains more than 90 percent uranium-235.

In 1941, a second element able to undergo a fission chain reaction was discovered, plutonium. Unlike uranium-235, plutonium does not exist in significant concentrations in nature. It is produced by the capture of neutrons by uranium-238 nuclei in a nuclear reactor. The first reactor was built under the leadership of Enrico Fermi at the University of Chicago in December 1942 as part of the Manhattan Project, and the first reactors designed to produce plutonium on a significant scale were built on the Columbia River in central Washington State. According to General Leslie Groves, the man in charge of the Manhattan Project, an isolated location for the production reactors was chosen because "no one knew what might happen, if anything, when a chain reaction was attempted in a large reactor." One fear was "some unknown and unanticipated factor" might lead a reactor "to explode and throw out great quantities of highly radioactive materials into the atmosphere."[6]

Separation of plutonium from neutron-irradiated uranium can be done chemically but requires a "reprocessing plant" in which all operations are performed behind thick concrete radiation shielding. Once the plutonium is separated from highly radioactive fission products, however, it can be handled relatively easily and could potentially be made into nuclear weapons, even by a subnational group. This is why proposals made to separate plutonium out of spent power reactor fuel for recycle in nuclear fuel have become so controversial.

The Manhattan Project developed different types of nuclear-weapon designs. A simple but very inefficient "gun-type" design was developed for highly enriched uranium and a much more difficult but efficient "implosion" design was developed for plutonium when it became apparent that the gun-type design would not work for this element. The Hiroshima bomb contained about 60 kilograms of highly enriched uranium while the Nagasaki bomb, which had about the same explosive energy, contained only 6 kilograms of plutonium. Modern versions of the Nagasaki design can use less than 4 kilograms of plutonium or 12 kilograms of highly enriched uranium.

Modern thermonuclear weapons use a "primary" fission explosion to trigger a much more powerful "secondary" explosion involving the fusion of the nuclei of heavy hydrogen atoms. These weapons generally contain both plutonium and highly enriched uranium—on average about 3–4 kilograms of plutonium in the fissile "pit" of the fission primary and 15–25 kilograms of highly enriched uranium in the fission-fusion secondary.

The current nine nuclear weapon states all followed different paths to weapons but in most cases relied on the example and even direct assistance of others. The histories of fissile material production for weapons by these countries are described in chapter 3. The first and most important case is that of the United States. The Manhattan Project, which included in its technical leadership a cadre of European émigré and refugee scientists, pioneered large-scale deployment of the technologies of uranium enrichment and plutonium production and provided a technological roadmap for most of the nuclear weapon programs that followed. The Soviet Union in particular patterned its first fissile material production facilities and its first weapon design on those of the United States. Later, in the early 1960s, the Soviet Union broke new ground by developing and deploying gas centrifuge uranium enrichment technology on a large scale.

Britain's nuclear weapons program was led by physicists who had participated in the U.S. wartime program. France's fissile material production complex, which was built in the early 1950s, followed Britain in its choice of technologies and scale. Outside information also was important for China, the first country to acquire nuclear weapons without an advanced scientific and industrial base. Many of China's nuclear experts were trained in the Soviet Union, which also provided expert advisors and designs for China's fissile material production facilities. The Soviet experts were withdrawn before China's uranium enrichment and plutonium production facilities were completed, however, and there were delays as China struggled to complete them.

Israel received secret assistance from France, which provided a complete plutonium production complex, both a nuclear reactor and a reprocessing plant. Faced with a uranium shortage, Israel also engaged in nuclear cooperation and trade with South Africa, receiving uranium for Israel's plutonium-production reactor.

India, the seventh state to acquire nuclear weapons, initially claimed to be interested only in nuclear power when it sought assistance in building nuclear facilities but made a point of keeping its options open. It eventually extracted plutonium from the fuel of a research reactor provided to it for peaceful purposes by Canada and the United States and used the plutonium to make nuclear weapons. For its part, Pakistan took advantage of the growing number of international civilian nuclear technology suppliers in the 1970s—especially those based in Europe—to purchase key components and materials for its gas centrifuge program. Pakistan also received direct assistance from China, including the design of a tested warhead.

In North Korea, the most recent state to have developed and tested nuclear weapons, the Yongbyon plutonium-production reactor is based on the published design of a 1950s era reactor developed by the United Kingdom. Its uranium enrichment program, revealed in 2010, is based on technology transferred from Pakistan in the 1990s. It is possible that North Korea has used this capability to make highly enriched uranium for weapons.

The final case is that of South Africa, which produced nuclear weapons of the gun-type design using highly enriched uranium produced by an inefficient aerodynamic uranium isotope separation process. Since South Africa gave up its nuclear weapons in 1991, the HEU recovered from them has been stored under international monitoring.

Chapter 4 provides an overview of the current global stockpile of roughly 1,900 tons of highly enriched uranium and plutonium categorized by current or intended use. These categories include fissile material in or committed to weapons, in naval nuclear propulsion programs, civilian material, and weapon material that has been declared excess for military purposes and is intended to be used in reactor fuel or disposed of in some other way.

The largest stockpiles both of HEU and plutonium are held by Russia, followed by those of the United States. There are significant uncertainties in estimates of most of the military stockpiles (typically of the order of 20–30 percent), since only the United States and the United Kingdom have so far made public declarations of their military HEU and plutonium inventories. The uncertainty in the estimate of Russia's stockpile of HEU is on the order of 100 tons

These estimates make clear, however, that most of the global fissile material stockpile is in military complexes and is overwhelmingly allocated for weapon purposes, with central estimates amounting to about 935 tons of HEU and almost 140 tons of plutonium. Because of the dismantlement of excess Cold War weapons by Russia and the United States, these stocks are far larger than needed for the actual current warhead stockpiles held by the nuclear weapon states.

A second stock of military fissile material is the almost 180 tons of HEU allocated to fuel naval propulsion reactors, mostly by the United States. Additionally, there are the roughly 260 tons of plutonium separated in several civilian nuclear energy programs. In terms of nuclear weapon equivalents, this plutonium stockpile by itself is sufficient for more than 30,000 nuclear warheads. Finally, about 60 tons of HEU are dedicated to civilian research reactors. While this is the smallest category of fissile material, it is spread across the largest number of states and has been the focus of the most strenuous efforts to make it more secure and to reduce the number of facilities in which it is stored.

The uncertainties in estimates of the global fissile material stockpile are equivalent to several thousand nuclear warheads. As nuclear arsenals are further reduced, uncertainties of such magnitude could become obstacles to progress toward nuclear disarmament. Greatly increased nuclear transparency, including the development of new bilateral and multilateral cooperative approaches to verify declarations by states of their fissile material production histories, will be required to reduce the uncertainties in estimated national fissile material inventories and build international confidence to support deeper cuts in the nuclear arsenals. In addition to an examination of production records, this verification will include techniques of nuclear archaeology, which involve the use of physical measurements at shutdown fissile material production facilities to independently estimate historical production at the facilities.

The Nuclear Weapons–Nuclear Energy Link

Along with the emergence of the nuclear weapon states, there was a spread of civilian nuclear technology to scores of countries, impelled in significant degree by the U.S. and Soviet Atoms for Peace initiatives launched in 1953. Atoms for Peace also led in 1957 to the establishment of the International Atomic Energy Agency (IAEA) with a mandate both to promote peaceful uses of nuclear technology and to monitor these uses to assure that fissile materials are not diverted to weapons use. This approach to managing the diversion risks of civilian nuclear energy programs was codified in the Treaty on the Non-Proliferation of Nuclear Weapons (commonly known as the Non-Proliferation Treaty or NPT) of 1970. Part II of this book addresses the challenges of stopping the spread of nuclear weapons to countries that do not have them and minimizing the risk of terrorist groups getting access to weapon-usable materials.

Almost any state could construct a nuclear device if it obtained the requisite amount of highly enriched uranium or separated plutonium or other less common fissile material. It is also prudent to assume that a terrorist group could make a nuclear explosive device if it had access to fissile material. A 1988 study by J. Carson Mark, for twenty-five years head of the Theoretical Division at the Los Alamos National Laboratory, and four colleagues, including the well-known nuclear weapons designer Theodore Taylor, concluded:

Crude nuclear weapons (similar to the Hiroshima gun-type and Nagasaki implosion-type weapons) could be constructed by a group not previously engaged in designing or building nuclear weapons provided that they have the technical knowledge, experience, and skills in relevant areas, e.g., the physical, chemical, metallurgical and nuclear properties of the various materials to be used, as well as the characteristics affecting their fabrication, and the technology of high explosives and/or chemical propellants.[7]

Efforts to strengthen the barriers to nuclear weapons production therefore must focus especially on increasing the difficulties a country or subnational group would face in acquiring fissile materials. The most plausible route to nuclear weapons for subnational groups would be to obtain separated plutonium or highly enriched uranium in fresh reactor fuel or from a fuel fabrication facility. Eliminating fuels that contain such materials therefore could greatly reduce the danger of terrorist acquisition of nuclear weapons.

With respect to nuclear weapon proliferation by countries, the situation is more complicated because countries could produce as well as divert fissile material. Speaking on the sidelines of the April 2012 Nuclear Security Summit in Seoul, South Korea, U.S. President Barack Obama observed: "The very process that gives us nuclear energy can also put nations and terrorists within the reach of nuclear weapons. We simply can't go on accumulating huge amounts of the very material, like separated plutonium, that we're trying to keep away from terrorists."[8]

Chapter 5 describes the spread of nuclear technology for civilian purposes initiated by the 1953 Atoms for Peace initiative.[9] The worldwide promotion of nuclear power that accompanied the Atoms for Peace programs stimulated interest in developing national nuclear science and engineering communities and industries. Atoms for Peace included the export, mostly by the United States and Soviet Union, of research reactors and highly enriched uranium fuel for them to over forty countries.

Inevitably, a growing number of countries sought to acquire the technologies to enrich uranium for light water reactor fuel and to separate plutonium to fuel plutonium breeder reactors, which the United States and other industrialized countries saw as the reactors of the future. The acquisition of reprocessing and enrichment technologies puts countries just a short step away from nuclear weapons, should they decide to seek them.

There are two principal nuclear "fuel cycles" in use. Most of the thirty or so countries with nuclear power plants, including the United States, which has about one-quarter of the world's power reactors, use fuel "once-through." The dominant reactor type, the light water reactor, so called because it is cooled by ordinary "light" water, is fueled by low-enriched uranium with about 3–5 percent uranium-235. The discharged spent fuel is stored pending its final disposal. This once-through fuel system has the critical advantage that nowhere in it is weapon-usable fissile material easily accessible. The low-enriched uranium in the fresh fuel cannot sustain an explosive chain reaction without further enrichment, and the plutonium in the spent fuel is never separated from the highly radioactive products of uranium fission. This effectively eliminates the chances of a subnational group acquiring fissile material from this fuel cycle.

The principal proliferation concern about countries using low-enriched uranium fuel once through is that a national enrichment plant designed to produce the low-enriched uranium could be converted rapidly to produce weapon-grade uranium. This possibility has been at the heart of international anxiety about Iran's uranium enrichment program, which has been producing material enriched to less than 5 percent uranium-235 for possible power reactor fuel as well as a smaller amount enriched almost to 20 percent uranium-235 for use in the Tehran Research Reactor.

The second prevalent nuclear fuel cycle is used primarily in France, which after the United States operates the second largest number of power reactors. It involves the separation of plutonium from the spent reactor fuel at a reprocessing plant and its recycle in reactor fuel, either as a mixed uranium plutonium oxide (MOX) fuel in light water reactors or as fuel for breeder reactors.

From the earliest days of the nuclear era, interest in civilian reprocessing was driven by the dream of breeder reactors that would produce more chain-reacting material than they consumed, typically by converting non-chain-reacting uranium-238 into plutonium. One legacy of this effort to commercialize breeder reactors is the global stockpile of 260 tons of civilian plutonium. Chapter 6 focuses on ending the separation of plutonium for reactor fuel.

Breeder reactors have been plagued by high capital costs and reliability problems. Despite over $100 billion spent over a period of fifty years in over half a dozen countries on breeder reactor research, development, and demonstration, commercialization efforts largely failed. Ambitious breeder development programs in the United States, the United Kingdom, and Germany were abandoned in the 1980s and 1990s. France and Japan have postponed their breeder programs. As of 2013, only Russia and India—recently joined by China on a pilot scale—separate plutonium with the intention of using it as fuel for prototype plutonium breeder reactors.

While the United States gave up both on breeder reactors and on reprocessing in the early 1980s, some other countries continued reprocessing even after postponing breeder commercialization and decided to mix their separated plutonium with uranium in MOX fuel for existing light water reactors. The spent MOX fuel is stored pending future developments. This fuel cycle is more costly than the once-through fuel cycle and also complicates radioactive waste disposal. Most countries that have tried it have therefore abandoned it. In 2012, the United Kingdom decided to end reprocessing when it completed its existing contracts. In France and Japan, as of this writing, the future of reprocessing was in debate but still unresolved.

In the 1970s, India demonstrated that civilian reprocessing opens a route to nuclear weapons for non-weapon states. It also creates a risk of the theft and use of plutonium by subnational groups since, unlike spent nuclear fuel assemblies, separated plutonium has no radiation barrier, making it easy to handle (e.g., in a simple glove box), and less than 8 kilograms are required to make a nuclear weapon. Plutonium oxide also could be used in a radiological dispersal device because it is extremely carcinogenic if inhaled.

Given that the breeder reactors that provided the original justification for the civilian separation of plutonium have proven largely unworkable, unnecessary, and uneconomical, and that separated plutonium is a clear proliferation risk, it would be appropriate to phase out the continued separation and use of plutonium as a fuel.

Chapter 7 assesses the challenges of ending the use of HEU as a reactor fuel. HEU is not used as a power reactor fuel but it is used to fuel a large fraction of the world's research and naval propulsion reactors. Access to 50–100 kilograms HEU would give even a relatively unsophisticated group the means to build a nuclear explosive device. Manhattan Project physicist and Nobel laureate Luis Alvarez observed that an improvised HEU nuclear explosive device with a possible yield of several kilotons could be as simple as an arrangement for dropping one subcritical mass of HEU onto another to create a supercritical mass.[10]

The U.S.-led Global Threat Reduction Initiative is carrying forward the effort to minimize civilian use of HEU in research reactors by converting them to low-enriched uranium fuel and removing fresh and spent HEU fuel from their sites. Russia has cooperated in this effort in countries to which the Soviet Union provided HEU-fueled research reactors. The U.S. and Russian efforts have successfully cleaned out HEU from about half the more than forty nonnuclear weapon states with HEU and are making progress in the other half. However, Russia has since 2010 started to give priority to converting or shutting down its own HEU-fueled research reactors; as of 2013, these accounted for more than half of the world's remaining unconverted facilities.

HEU in naval fuel cycles also is a security risk. The single largest illicit diversion of fissile material may have been the hundreds of kilograms of weapon-grade uranium that were secretly transferred in the 1960s from a naval fuel fabrication facility in the United States to Israel, apparently with the cooperation of the plant's owner (see chapter 3). In 1993, the theft of a much smaller amount of HEU submarine fuel from a Russian storage facility helped focus attention on the need to secure Russia's nuclear materials after the collapse of the Soviet Union.[11]

Despite the greater attention given to fissile material security after the attacks of September 11, 2001, significant problems with protecting HEU in storage have surfaced even in the U.S. nuclear weapons complex. In July 2012, three antinuclear activists successfully penetrated the high security system surrounding the newly built Highly Enriched Uranium Materials Facility at the Y-12 site in Tennessee, which contains over 100 tons of highly enriched uranium.[12] The breach was attributed to a range of problems, including both inoperable and poorly functioning security cameras, a failure to react to alarms, and inadequate response plans and procedures at the site. The United States Department of Energy, which manages the nuclear weapons complex, spends about $1 billion a year on physical security of nuclear materials.

Large quantities of HEU are used to fuel the nuclear navies of the United States, Russia, and the United Kingdom, and India planned to begin trials in 2014 of its first HEU-fueled nuclear submarine. The United States alone has reserved 152 tons of weapon-grade uranium for future use in naval propulsion reactors—enough for 5,000–10,000 nuclear explosives. France, however, fuels its submarines and a nuclear-powered aircraft carrier with low-enriched uranium (LEU) fuel. It is believed that China also uses LEU fuel and Brazil, which is planning to be the first non-weapon state to have a nuclear-powered submarine, also has chosen to use LEU fuel—at least initially. If the countries that now use HEU in naval fuel also converted to LEU, much more HEU could be declared excess and eliminated. But this issue has thus far not received much attention.

Eliminating Fissile Materials

The third and final part of this book (chapters 8, 9, and 10) explores how initiatives to cap and reduce fissile material stockpiles could support and even drive progress toward nuclear disarmament. A nuclear weapon-free world would be far more stable if it were also a fissile material-free world. This would involve ending production of fissile materials and eliminating the current global fissile material stockpiles in a verifiable and irreversible manner.

Chapter 8 discusses a proposed treaty that would end the production of fissile material for weapons, the Fissile Material Cutoff Treaty (FMCT). Such a treaty would build on the fact that the United States, Russia, the

United Kingdom, and France have already declared that they have permanently ended their production of fissile materials for weapons, while China has suspended production. An FMCT has been under consideration at the United Nations Conference on Disarmament in Geneva since 1993, but linkages to other issues—especially by Pakistan—have blocked the launch of formal negotiations.

Once negotiations do go forward, the two critical issues will be the scope of the treaty and how it will be verified. The principal question on scope is whether the weapon states will commit permanently and verifiably not to use for weapons preexisting stocks of fissile material that they currently consider excess to their weapons needs.

Much of the verification of an FMCT could be carried out using the same monitoring techniques developed by the IAEA to verify that non-weapon states are complying with their NPT obligation not to use fissile materials to produce "nuclear weapons or other nuclear explosive devices." The IAEA or (less likely) another inspection agency established by the treaty would monitor enrichment plants in the weapon states to determine whether they are producing HEU and, if they are, monitor the subsequent storage or use of the HEU. Plutonium and other specified fissile materials newly separated at weapon state reprocessing plants would similarly be subject to such international monitoring. And, if it were part of the agreement, the storage and use of preexisting fissile materials that the weapon states have declared excess for all military purposes too would be monitored.

It will be critical under an FMCT to verify that there is no clandestine production of highly enriched uranium or plutonium at undeclared sites. To accomplish this, international inspectors would have to be authorized to take environmental samples within weapon states as well as non-weapon states and to initiate "challenge inspections" at suspect facilities. Analysis of air samples might, for example, be able to detect degradation products of uranium hexafluoride leaked from enrichment operations or elevated levels of krypton-85, a gaseous fission product released by reprocessing.

The most effective and enduring way to deal with the security dangers posed by fissile materials is to dispose of them as irreversibly as possible. Chapter 9 describes current approaches to disposing of highly enriched uranium and alternative options for disposing of plutonium.

The disposal of highly enriched uranium is relatively straightforward. It is blended with natural or slightly enriched uranium to produce low-enriched uranium that can then be used as a fuel for light water nuclear power reactors commonly used to produce electricity. As of 2013, Russia and the United States have already down-blended a combined total of over 600 tons of highly enriched uranium, most dramatically with 500 tons of Russian weapon-grade uranium blended down for fuel for U.S. reactors under a U.S.-Russian deal concluded in 1993.

The disposal of weapons plutonium has proven costly and complicated and has made little progress. Although the United States and Russia concluded in 2000 a Plutonium Management and Disposition Agreement that committed each party to dispose of at least 34 tons of weapon-grade plutonium "withdrawn from nuclear weapon programs," this process is currently not scheduled to begin at least until 2018. The United States has planned to fabricate most of its excess plutonium into MOX fuel for use in light water reactors but, as of 2013, the future of this program was in doubt.

Irradiating plutonium in a light water reactor as MOX fuel eliminates a fraction of the initial material. The remaining plutonium survives but becomes encapsulated in a highly radioactive spent fuel matrix similar to the one from which it was first separated. The plutonium could not be used for weapons without once again separating it from the spent fuel in a reprocessing plant. The spent MOX fuel would be disposed of in a geological repository. As MOX program costs have spiraled and delays mounted, the United States has decided to consider alternative means of disposal of its excess weapons plutonium.

Russia plans to use its excess separated plutonium in prototype breeder reactors. As with the MOX option, the spent fuel still contains plutonium. Russia intends to eventually separate the plutonium and use it again as reactor fuel, creating in effect a perpetual river of separated plutonium.

Nonreactor options for plutonium disposal merit greater attention than they have received. One is to mix the plutonium back into the concentrated fission-product wastes from which it was originally separated. This plutonium-bearing waste could then be placed in a deep geological repository with regular spent nuclear fuel. Another option would be to immobilize the plutonium in a durable matrix and dispose of it underground in 3–5 kilometer deep boreholes.

Chapter 10 offers a summary of the fissile material perspective on nuclear disarmament and makes a case for three policy goals that could reduce the danger from the large global stock of fissile material and nuclear weapons.

1. Radically increase the transparency of nuclear warhead and fissile material stockpiles to allow for verified deep reductions in these stockpiles as part of the process of nuclear disarmament.

2. Verifiably end all further production and use of HEU and plutonium as fuel both for civilian and military reactors.

3. Verifiably dispose of military and civilian fissile material stockpiles as irreversibly as possible.

Confidence in and verification of nuclear disarmament will be far easier in a world where there is no production or use of separated plutonium or highly enriched uranium and where fissile material stocks have been eliminated. More drastically, if civilian nuclear power were phased out in parallel with nuclear weapons, it would become more difficult and more time-consuming for any country to make fissile materials for nuclear weapons and would make it easier for the international community to detect and respond to what would be a clear threat to international peace and security.

I

How the Nuclear World Emerged

2

Production, Uses, and Stocks of Fissile Materials

The fissile materials used in nuclear weapons are plutonium and highly enriched uranium, uranium that contains at least 20 percent of the isotope uranium-235. This chapter lays out how these fissile materials are produced and how they are used in nuclear weapons.

What Are Fissile Materials?

Nuclear fission was discovered in December 1938 by Otto Hahn and Fritz Strassmann in Berlin, and explained soon afterward by Lise Meitner and her nephew, Otto Frisch, then refugees in Stockholm and Copenhagen, respectively. Ever since Henri Becquerel and Marie and Pierre Curie discovered radioactivity around the turn of the century, scientists had dreamed of a nuclear reaction that would release energy for practical use. Frederick Soddy, the British radiochemist and Nobel Laureate, also saw the danger from nuclear energy and issued the following warning in 1903:

It is probable that all heavy matter possesses—latent and bound up with the structure of the atom—a similar quantity of energy to that possessed by radium. If it could be tapped and controlled, what an agent it would be in shaping the world's destiny! The man who put his hand on the lever by which a parsimonious nature regulates so jealously the output of this store of energy would possess a weapon by which he could destroy the earth if he chose.[1]

The writings of Soddy and other contemporary scientists influenced H. G. Wells, who offered his vision of nuclear war in his 1914 novel, *The World Set Free*.[2] Wells's imaginings in turn inspired Hungarian physicist Leo Szilard in 1933 to conceive of the possibility of a nuclear chain reaction. The following year, Szilard, then a refugee living in England, submitted a secret patent application for a nuclear chain reaction.[3] Szilard correctly identified neutrons, which had just been discovered, as

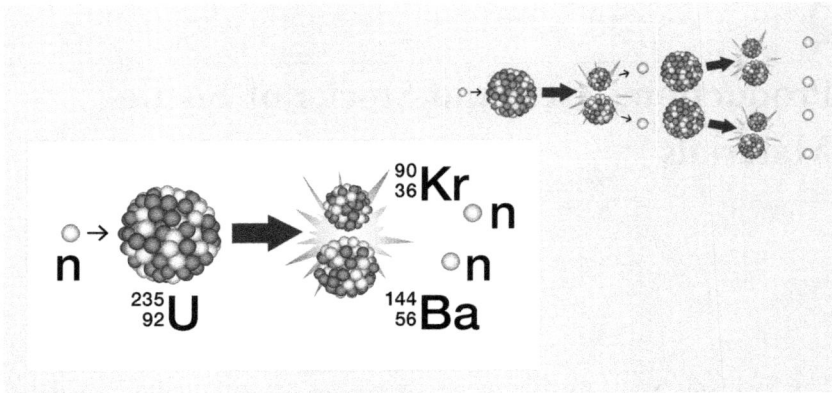

Figure 2.1

An explosive fission chain reaction releases enormous amounts of energy in one-millionth of a second. In this example, a neutron is absorbed by the nucleus of uranium-235, which splits into two fission products (in this case, barium and krypton). The released energy is carried mainly by the fission products, which repel each other and separate at high velocities, colliding with and heating the atoms in any surrounding matter. Additional neutrons are released in the process, which therefore can set off a chain reaction in a critical mass of fissile material. The chain reaction proceeds very rapidly; in a supercritical mass of pure HEU or plutonium, eighty doublings of the neutron population in a millionth of a second can fission one kilogram of material and release an energy equivalent to 18,000 tons of high explosive (TNT).

the subatomic particles that would maintain the chain reaction. However, the particular reaction process he envisioned (involving beryllium) would ultimately prove infeasible.

Once fission was discovered, an avalanche of research followed in several countries: in 1939 alone, there were nearly one hundred articles revealing further details of the fission process. By the end of the year, a basic understanding was in hand. When the nucleus of a heavy fissile atom absorbs a neutron, it will usually split into two medium-weight nuclei, but to sustain a chain reaction each fission event also has to release on average more than one neutron able to cause additional fissions (figure 2.1). That fissions of uranium release two to three neutrons on average was confirmed in March 1939 by experiments in Paris by Frédéric Joliot-Curie, Lew Kowarski, and Hans von Halban, and independently by two groups in New York led by Italian physicist Enrico Fermi and by Leo Szilard.

The fission of a single nucleus yields one hundred million times more energy per atom than the burning of a molecule of hydrogen, and the time between two generations in a chain reaction can be extremely short, resulting in the possibility of an explosive release of energy.

The implications of these discoveries were obvious to many nuclear scientists. In August 1939, Leo Szilard and Albert Einstein drafted, and Einstein signed, a now-famous letter informing U.S. President Roosevelt of the following:

In the course of the last four months it has been made probable—through the work of Joliot in France as well as Fermi and Szilard in America—that it may become possible to set up a nuclear chain reaction in a large mass of uranium, by which vast amounts of power and large quantities of new radium-like elements would be generated. Now it appears almost certain that this could be achieved in the immediate future. . . . It is conceivable—though much less certain—that extremely powerful bombs of a new type may thus be constructed.[4]

Some researchers were contemplating how a nuclear bomb might be constructed. Most notably, in 1940 at the University of Birmingham in England, Otto Frisch, now a refugee in Britain, and another German physicist refugee, Rudolph Peierls, wrote a pathbreaking memorandum on the possibility of a "super bomb." The memorandum built on an insight of the Danish physicist, Niels Bohr, and American physicist John Wheeler. In 1939, the two had identified uranium-235 as the natural uranium isotope that can sustain a fission chain reaction and had explained the physics of fission.[5] Natural uranium contains only 0.7 percent uranium-235. The remainder is almost entirely uranium-238 (99.3 percent).[6]

Frisch and Peierls therefore focused on the possibility of a bomb based on uranium-235 and noted the existence of various isotope-separation techniques, which later came to be known as uranium "enrichment" and could, in principle, allow the extraction of nearly pure uranium-235 from natural uranium.[7] They attempted to determine the "critical size" (now termed critical mass) of uranium-235 required to sustain an explosive chain reaction. They estimated the critical mass of pure uranium-235 to be 0.5 kilograms and that a "suitable size" of the amount for a bomb would be 1 kilogram. These estimates were too low by a factor of about one hundred, in part due to a high estimate of the neutron "cross section" used in their calculations. Perhaps in part because of this underestimate of the magnitude of the task, the memorandum galvanized research on nuclear weapons in the United States as well as the United Kingdom.[8] In Germany, Werner Heisenberg may have overestimated the magnitude of a critical mass of uranium-235 by a factor of ten and concluded that the production of a nuclear weapon would not be possible before the end of World War II.[9]

Initial efforts in the British and U.S. nuclear weapon programs focused on developing processes to separate fissile uranium-235 from the much more abundant non-fissile uranium-238 isotope on a sufficiently large scale.

By the definition used in this book, fissile materials are those that can sustain an explosive fission chain reaction.[10] Uranium with a uranium-235 content of 20 percent or higher, the definition of highly enriched uranium, is considered weapon-usable. Uranium enriched to 90 percent uranium-235 and higher is often called weapon-grade.

Uranium also can be used in a reactor to produce a controlled non-explosive chain reaction with a steady rate of energy release. In the power reactors deployed around the world, this is accomplished by distributing the uranium in a "moderating" medium that slows the neutrons between fissions without absorbing them. This sharply increases the probability of fission absorption in uranium-235 relative to non-fission absorption by uranium-238 or other elements. Neutron moderators include graphite, ordinary "light" water (H_2O), and heavy water (D_2O). With pure graphite or heavy water moderators, which absorb very few neutrons, natural uranium can be used as a reactor fuel. Most power reactors use light water as moderator, but, because of neutron absorption by the water, require the use of low-enriched uranium, which typically contains 4–5 percent uranium-235.

A second fissile material was soon discovered. Nuclear experiments had been carried out since the mid-1930s in which natural uranium, exposed to neutrons, produced new isotopes and elements when uranium, primarily uranium-238, captured one or more neutrons without fissioning. One of the first elements discovered in this way was plutonium-239, a radioactive decay product of the short-lived isotope uranium-239. That plutonium-239 can undergo fission was confirmed in the spring of 1941, shortly after the element was first isolated.[11]

Other man-made isotopes—including uranium-233, neptunium-237, and americium-241—also can sustain a chain reaction but, as far as is publicly known, only uranium-235, plutonium, and uranium-233 have been used in nuclear weapons.[12]

Among the alternative nuclear materials, uranium-233 and neptunium-237 may pose additional future verification challenges because significant stockpiles exist in the United States, Russia, and perhaps other nuclear weapon states. Neptunium-237 can be separated from spent reactor fuel and used in irradiation targets to produce plutonium-238, which is used in thermoelectric generators for space and other applications. The global inventory of separated neptunium could be on the order of one ton.[13]

The amount of material required for a critical mass can vary widely depending on the material, its chemical form, and the characteristics of the surrounding materials that reflect neutrons back into the material. The unreflected ("bare") and reflected critical masses of some fissile materials are shown in figure 2.2.

Kilograms (kg)

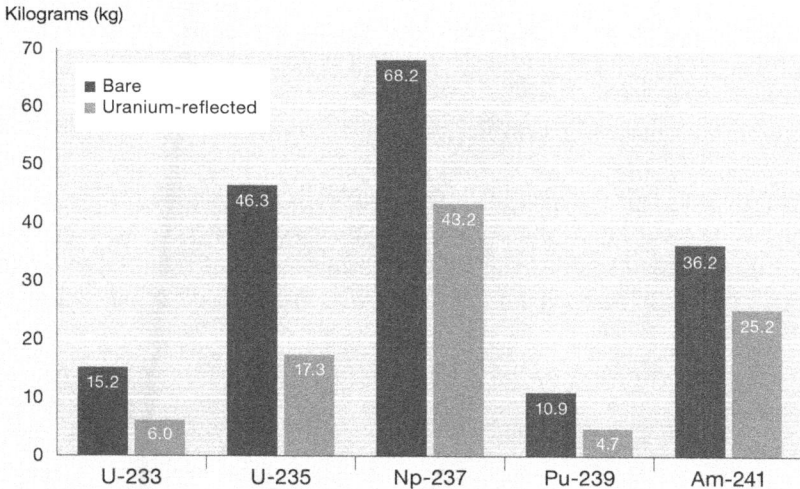

Figure 2.2

Unreflected (bare) and reflected critical masses for key fissile isotopes. A bare critical mass is the spherical mass of fissile metal barely large enough to sustain a fission chain reaction in the absence of any material around it. Neutron-reflecting materials can be used to reduce the critical mass substantially, in this case a 10-cm thick shell of natural uranium. Uranium-233, neptunium-237, and americium-241 are, like plutonium-239, reactor-made fissile isotopes and could potentially also be used to make nuclear weapons but have apparently not been used to make other than experimental devices. These values, determined from MCNP5 calculations, represent idealized cases; actual fissile materials will contain mixtures of isotopes. For example, U.S. weapon-grade uranium contains 1 percent uranium-234, 93 percent uranium-235, and 6 percent uranium-238.

Production of Fissile Materials

Fissile materials do not occur in nature. They must be produced through complex physical and chemical processes. In the early 1940s, before the U.S. Manhattan Project was launched, it was impossible to anticipate just how difficult it would be to make fissile materials, to enrich uranium or to build and operate a reactor and separate plutonium from the irradiated uranium. Production of fissile materials turned out to be the major barrier to the acquisition of nuclear weapons, requiring a dedicated and large industrial infrastructure. After touring the Manhattan Project sites, Niels Bohr was quoted as saying: "I told you it couldn't be done without turning the whole country into a factory. You have done just that."[14] The difficulties associated with producing these materials are the main technical barriers to the acquisition of nuclear weapons.

Highly Enriched Uranium

Uranium-235 makes up only 0.7 percent of natural uranium (figure 2.3). Although an infinite mass of uranium with a uranium-235 content of 6 percent could, in principle, sustain an explosive chain reaction, weapons experts have advised the International Atomic Energy Agency that uranium enriched to above 20 percent uranium-235 is required to make a fission weapon of practical size. The IAEA therefore describes uranium

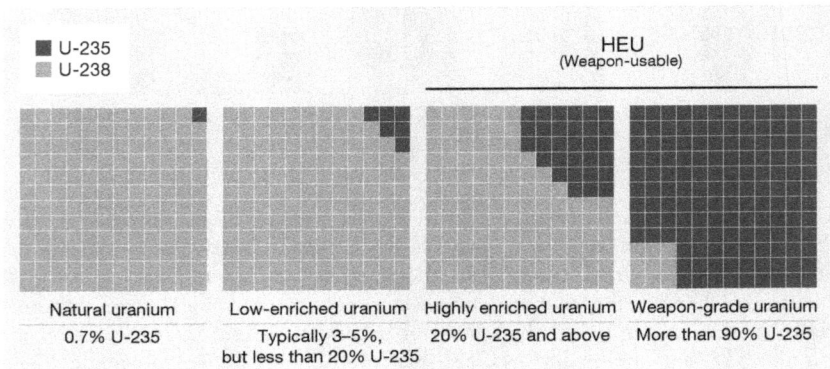

■ U-235		HEU
■ U-238		(Weapon-usable)

Natural uranium	Low-enriched uranium	Highly enriched uranium	Weapon-grade uranium
0.7% U-235	Typically 3–5%, but less than 20% U-235	20% U-235 and above	More than 90% U-235

Figure 2.3

Natural and enriched uranium. Below 20 percent enrichment, uranium is defined as low-enriched uranium and, in contrast to highly enriched uranium, considered non-weapon-usable. Weapon states typically have enriched uranium to 90 percent or more in order to minimize the size of nuclear warheads.

enriched to 20 percent or more as "highly enriched uranium" and considers it "direct use" weapon material. The critical mass of uranium drops rapidly as the enrichment level increases. To minimize their size, actual weapons typically use uranium enriched to 90 percent or higher. Such uranium is termed "weapon-grade."

Before it can be enriched, uranium has to be converted into a chemical form suitable for the chosen enrichment process.[15] The nuclear fuel cycle typically begins with uranium ore being mined and concentrated. Concentration of the uranium ore is carried out at a mill or in situ by chemical leaching, which increases the uranium oxide content from the typical fraction of 0.1 percent in raw ore to 85–95 percent. This requires processing large amounts of ore. Uranium mills are typically close to the mines in order to minimize transport costs. The product of the mill is yellowcake, U_3O_8, a bright yellow powder that is widely traded as a commodity. The yellowcake is then further purified and processed at a conversion plant. Typically, conversion plants produce uranium hexafluoride (UF_6) for enrichment plants, but other outputs are possible, such as natural uranium dioxide (UO_2) or uranium metal for use in reactor fuel. Production of sufficiently pure UF_6 for enrichment or of "nuclear-grade" uranium oxide or metal for reactor fuel is technically difficult. IAEA safeguards on uranium typically begin at conversion plants.

To produce uranium enriched in uranium-235 requires sophisticated isotope separation technology since the isotopes uranium-235 and uranium-238 are chemically virtually identical and differ in weight by only about 1 percent. Although uranium enrichment technology is widely understood, facilities that can produce enriched uranium on a scale sufficient to make nuclear weapons or to sustain a light water power reactor are found in only a relatively small number of nations (appendix 1).

In a uranium enrichment facility, the feed stream, usually natural uranium, is split into two streams: a product stream enriched in uranium-235 and a waste or "tails" stream depleted in uranium-235. The work of isotope separation is measured in separative work units (SWU). (There are two SWU units: kg-SWU and ton-SWU [1,000 kg-SWU]. In this book, SWU refers to kg-SWU.) The capacity of commercial enrichment facilities is commonly measured in millions of SWU per year (SWU/yr).

Historically, several technologies have been used to separate uranium isotopes. The first method considered was thermal diffusion, invented by German chemists Klaus Clusius and Gerhard Dickel in 1938.[16] In their 1940 memorandum, Frisch and Peierls cited this method as possibly the only one that could be scaled up to produce the quantities of HEU needed for a nuclear weapon. It involves placing a gas or liquid containing a mixture of isotopes into a very long vertical tube with a heated wire running along its axis and cooled walls. The lighter isotope has a slightly higher concentration near the hot wire and the heated, slightly enriched material rises while the depleted cooled material sinks. This "counter-current flow" multiplies the enrichment effect.

The United States built a production scale facility, S-50, at the Manhattan Project site at Oak Ridge, Tennessee, in about one year, starting in June 1944, using over 2,100 separation columns, each almost 15 meters high, filled with uranium hexafluoride.[17] The facility was used to produce slightly enriched uranium containing up to 1.4 percent uranium-235. It was shut down in September 1945.

A second enrichment method used electromagnetic isotope separation (EMIS). This process introduces a beam of uranium-containing ions into a magnetic field that splits the beam into two by virtue of the fact that the paths of the electrically charged ions containing the lighter uranium-235 atoms are bent more by the magnetic field. This method was implemented during the U.S. Manhattan Project using "calutrons" developed by Ernest Lawrence at the University of California, Berkeley. Electromagnetic separation was used for a time to enrich to over 80 percent the uranium product of the partially completed gaseous diffusion plant described below.[18] The Soviet Union used the method in a similar way,[19] and Iraq in the 1980s also tried to use electromagnetic separation in its nuclear weapons program.[20] The Iraqi program remained undetected for some time because few expected any country would use such an inefficient (and energy-intensive) process after other more efficient enrichment methods—notably gas centrifuges—had been successfully developed.

Gaseous diffusion was used to produce most of the highly enriched uranium in the world. Nobel laureate Gustav Hertz first effectively used the gaseous diffusion method for isotope separation in 1932 at the Siemens company in Berlin on a laboratory scale in experiments employing neon gas passed through a series of porous clay tubes. The method was applied to uranium enrichment on a laboratory scale in the UK World War II nuclear weapons program by a group led by Francis Simon, a refugee German physicist based at the Clarendon Laboratory at Oxford.[21] Gaseous diffusion was deployed on a large scale during the Manhattan Project at the K-25 plant at the Oak Ridge site. The Soviets captured Hertz at the end of World War II and took him to the Soviet

Union where he contributed to the development of gaseous diffusion technology for the Soviet nuclear weapons program.[22]

Gaseous diffusion isotope enrichment exploits the fact that, in a uranium-containing gas, the lighter molecules containing uranium-235 diffuse more quickly through the pores in a barrier than those containing uranium-238. The effect is only a few tenths of a percent, however, and the molecules have to be pumped through thousands of barriers to reach weapon-grade. This requires enormous amounts of energy. For example, France's Georges Besse gaseous diffusion plant (GDP) was powered by three dedicated 900-megawatt nuclear power plants. The GDPs in China, France, Russia, the United Kingdom, and the United States have all been closed down, with the last of the U.S. GDPs shut down in 2013. More economical gas centrifuge plants have replaced all of these plants.

All of those countries that have built new enrichment plants since the 1980s have chosen centrifuge technology. This was one of the earliest enrichment methods considered, with suggestions as early as 1919 that centrifuges might be the "most promising method" for separating isotopes.[23] The gas centrifuge was first successfully used for separating isotopes in 1934 by Jesse Beams at the University of Virginia who worked with chlorine gas. The United States sought to develop the Beams centrifuge for uranium enrichment during the Manhattan Project, along with electromagnetic separation, thermal diffusion, and gaseous diffusion, but eventually abandoned the effort in November 1942.[24] The Soviet Union also had a centrifuge program, but this did not mature until the late 1950s. Among the people who worked on this program was captured Austrian physicist Gernot Zippe, who would later contribute to the centrifuge programs in Germany and the United States and indirectly to the programs in a number of other countries.[25] (For further discussion of Zippe and of the spread of the centrifuge, see chapter 5.)

Modern gas centrifuges spin uranium hexafluoride (UF_6) gas at enormous speeds so that the uranium is pressed against the outer wall with more than 100,000 times the force of gravity. The molecules containing the heavier uranium-238 atoms concentrate slightly more toward the wall relative to the molecules containing the lighter uranium-235. An axial circulation of the UF_6 is induced within the centrifuge, which multiplies this separation along the length of the centrifuge and increases the overall separation by the machine greatly (figure 2.4). Since both the throughput and enrichment of a single machine are small, the process is

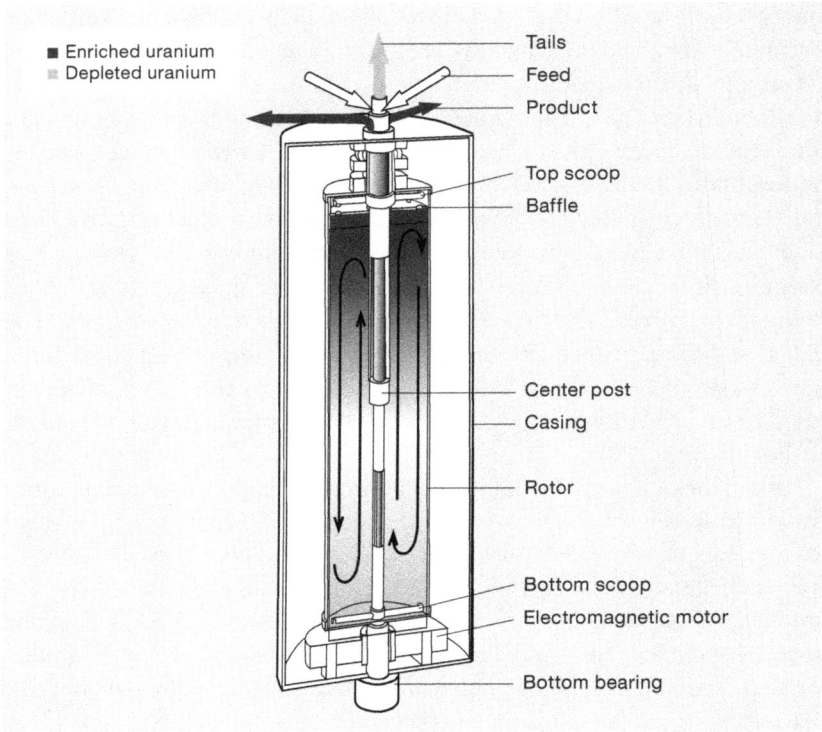

Figure 2.4

The gas centrifuge for uranium enrichment. The possibility of using centrifuges to separate isotopes was raised in 1919, shortly after isotopes were discovered. The first experiments using centrifuges to separate isotopes of uranium (and other elements) were successfully carried out on a small scale prior to and during World War II, but the technology only became economically competitive in the 1970s. Centrifuges are the most economical enrichment technology but also proliferation-prone. Machine capacities have increased from about 1–2 SWU/yr for a first-generation design to 50–100 SWU/yr for a modern centrifuge manufactured by Enrichment Technology Company.
Source: Graphics adapted from *Scientific American.*

repeated in a "cascade" of ten or more stages to produce uranium enriched to the 3–5 percent level used in most nuclear power reactors. More stages or a system of interconnected cascades are needed to produce weapon-grade HEU.

From a nonproliferation perspective, centrifuge technology has two major disadvantages relative to gaseous diffusion technology. First, the inventory held up in a typical gaseous diffusion cascade is about 1,000 tons as compared to a few kilograms in a centrifuge plant.[26] This means that it would take only hours to flush the uranium out of a centrifuge cascade and start feeding in enriched uranium to achieve higher enrichment levels.[27] This makes possible a "breakout" scenario, where a civilian enrichment plant is quickly converted to weapon use.[28]

Second, clandestine centrifuge facilities are extremely difficult to detect with remote-sensing techniques. A centrifuge plant with capacity to make HEU sufficient for a bomb or two per year could be small and indistinguishable from many other industrial buildings. Due to its low power consumption, there are no unusual thermal signatures as compared to other types of factories with comparable floor areas. Leakage of UF_6 to the atmosphere from centrifuge facilities is minimal because the gas in the pipes is below atmospheric pressure and air therefore leaks into the centrifuges rather than the UF_6 leaking out.

The capability of a centrifuge plant to convert relatively quickly from producing low-enriched uranium to weapon-grade uranium is especially problematic. To produce 20 tons of 4.3 percent uranium-235 (low-enriched uranium) for the annual reload of a typical 1,000-megawatt power reactor takes about 200 metric tons of natural uranium feed and 120,000 SWU. To produce 25 kilograms of weapon-grade uranium (93 percent U-235)—enough for a nuclear weapon (see below)—only takes about 5.5 tons of natural uranium feed and 5,000 SWU.[29] An enrichment plant such as the one that Iran has been building at Natanz, originally sized to support a single nuclear power plant, could therefore be adapted to produce enough weapon-grade uranium for about twenty nuclear weapons a year. If a stockpile of 4.3 percent enriched uranium were fed into it, the plant could hypothetically produce enough weapon-grade uranium for eight bombs a month.

Plutonium Production and Separation
Plutonium was formed in the same cosmic event that created the earth's uranium but does not occur naturally because its half-life is short compared to the age of the earth (table 2.1). It is produced in nuclear

Table 2.1
Key properties of plutonium isotopes and americium-241, which is a decay product of plutonium-241

	Half-life	Spontaneous fission rate	Neutron rate	Heat rate	WPu*	RPu*
Pu-238	88 years	1,180,000 fis/(kg s)	2,670,000 n/(kg s)	567 W/kg	0.05%	1.80%
Pu-239	24,110 years	7.1 fis/(kg s)	16.5 n/(kg s)	1.9 W/kg	93.60%	59.00%
Pu-240	6,570 years	479,000 fis/(kg s)	1,030,000 n/(kg s)	7.0 W/kg	6.00%	23.00%
Pu-241	14.4 years	<10 fis/(kg s)	<25 n/(kg s)	3.3 W/kg	<0.30%	<12.20%
Pu-242	374,000 years	805,000 fis/(kg s)	1,720,000 n/(kg s)	0.1 W/kg	0.05%	4.00%
Am-241	433 years	545 fis/(kg s)	1,360 n/(kg s)	115 W/kg	<0.30%	**
Weapon-grade plutonium		30,000 fis/(kg s)	64,000 n/(kg s)	2.5 W/kg		
Reactor-grade plutonium		164,000 fis/(kg s)	354,000 n/(kg s)	13.3 W/kg		

Source: www.nucleonica.net.
Notes: *WPu is a notional composition of weapon-grade plutonium and RPu of reactor-grade plutonium for a burnup of 40 thermal MW days/kg. **Americium-241 content in reactor-grade plutonium strongly depends on the age of the material and can be several percent. Fissions per kilogram and second (fis/kg s); neutrons per kilogram and second (n/kg s); watts per kilogram (W/kg).

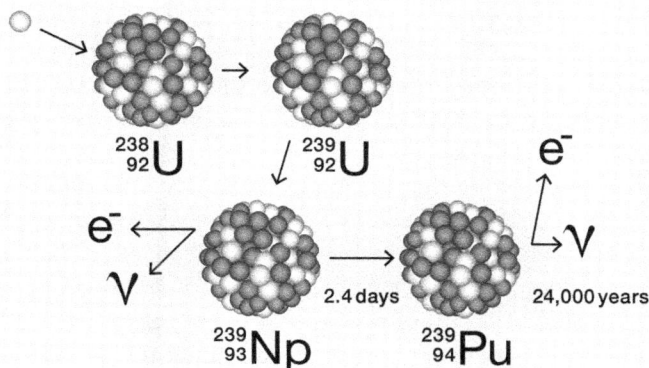

Figure 2.5

Making plutonium in a nuclear reactor. A neutron released by the fissioning of a chain-reacting uranium-235 nucleus is absorbed by a uranium-238 nucleus. The resulting uranium-239 nucleus decays with a half-life of 24 minutes into neptunium-239, which in turn decays into plutonium-239. Each decay is accompanied by the emission of an uncharged, massless neutrino and a negatively charged electron to balance the increase in the charge of the nucleus.

reactors when a uranium-238 nucleus absorbs a neutron creating short-lived uranium-239, which subsequently decays to neptunium-239 and then to plutonium-239 (figure 2.5).

Almost all reactors dedicated to the production of plutonium for weapons have been fueled with natural uranium. High-purity graphite or heavy water is used as the moderator material to slow the neutrons to sustain the chain reaction, and the core is cooled by water, air, or carbon dioxide.

The first plutonium production reactors were three U.S. reactors that started up in late 1944 and early 1945.[30] They were large cubes of graphite, about 10 meters on each side, made up of 100,000 separate blocks and weighing on the order of 2,000 tons. Pipes ran through over 2,000 horizontal channels in the graphite to hold about 100 tons of fuel in the form of cylindrical aluminum-clad metal slugs of natural uranium that both sustained the chain reaction and contained the uranium-238 that surplus neutrons converted into plutonium. The pipes also carried cooling water to remove the fission heat.

This design was copied by the Soviet Union and later by China. Britain and France also copied it, but with a carbon dioxide coolant after Britain's experiment with air cooling resulted in the Windscale fire of 1957. North Korea copied and downsized the UK designed gas-cooled graphite reactor for the reactor with which it produced weapon-grade plutonium.

The other reactor type that has been used to produce plutonium for weapons is the heavy water reactor first developed by Canada.[31] Canada's first heavy water reactor, the NRX at the Chalk River Laboratories in Ontario, was designed as a part of the U.S.-Canadian-UK World War II cooperative nuclear weapons program. It began operating at 20 megawatts in 1947 and provided relatively small quantities of plutonium to the U.S. postwar nuclear weapons program until 1950, after which it served as a research reactor.

The design of the reactor was similar to that of graphite reactors except that the graphite was replaced by an upright cylindrical tank of heavy water with a height of over 3 meters and a diameter of about 2.5 meters. The tank was sheathed in a graphite neutron reflector to reduce neutron loss from the core. About 200 vertical pipes running through the heavy water tank held 10 tons of aluminum-clad natural uranium metal fuel and carried cooling water. In some variants of this reactor design, heavy water is used as both moderator and coolant to reduce neutron losses and in others, light water is used as a coolant. The United States, Russia, France, Israel, India, and Pakistan all have used heavy water reactors for plutonium production.

In dedicated production reactors, about 0.8–1.0 grams of plutonium are produced per gram of uranium-235 fissioned or, almost equivalently, per thermal megawatt day of fission heat released. For example, India's CIRUS plutonium-production reactor, which was used to produce the plutonium for India's first nuclear weapons, had a thermal power of 40 megawatts and annually discharged about 10–12 tons of spent fuel containing 9–10 kilograms (about two bombs' worth) of weapon-grade plutonium. This estimate assumes a full-power operation 80 percent of the time and a discharge burnup of 1,000 thermal megawatt days/ton, achieved by limiting the residence time of the fuel in the reactor.

Plutonium also is produced in civilian power reactors, where the residence time of the fuel is typically much longer. The longer an atom of plutonium-239 stays in a reactor after it has been created, the greater the likelihood that it will absorb a second neutron and fission or become plutonium-240, which can absorb additional neutrons and become plutonium-241 or plutonium-242.[32] Plutonium therefore comes in a variety of isotopic mixtures. The plutonium in typical power-reactor spent fuel (reactor-grade plutonium) contains 50–60 percent plutonium-239 and about 25 percent plutonium-240. Weapon designers prefer to work with a mixture that is as rich in plutonium-239 as possible because of its relatively low rate of radioactive heat generation and relatively low spontaneous emissions of neutrons and gamma rays (table 2.1). "Weapon-grade" plutonium therefore contains more than 90 percent plutonium-239. As explained below, however, reactor-grade plutonium can sustain an explosive chain reaction, with a critical mass that is only about one-third greater than that of weapon-grade plutonium.

In light water power reactors, the net production of plutonium is only 0.2–0.3 grams of plutonium per thermal megawatt-day because about two-thirds of the plutonium is fissioned during the long residency of the fuel in the reactor core. A standard light water reactor with a power output of 1,000 megawatts electrical therefore produces about 250 kilograms of plutonium per year in its spent fuel. In the heavy-water-moderated CANDU power reactor, plutonium production per megawatt-day is about twice as large and the fraction in the heavier plutonium isotopes is smaller. CANDU reactors, which are fueled by natural uranium, are continuously refueled instead of a third of a core once every 12–18 months, which is typical for light water reactors.

Thus, uranium-based spent fuel from all types of reactors contains substantial amounts of plutonium. About 1 percent of the uranium loaded into a light water power reactor, for example, is discharged as plutonium. As long as the plutonium remains embedded in the spent fuel along with the highly radioactive fission products, however, it is relatively inaccessible. Due to its very intense gamma radiation field, spent fuel can only be handled from behind heavy shielding, which makes its diversion or theft by subnational groups unrealistic.

Separated plutonium can be handled without radiation shielding, however. It is dangerous primarily when inhaled—such as when plutonium metal burns or is already in an oxide powder form—and therefore must be handled in sealed containers or glove boxes.

Plutonium is separated from irradiated uranium in a "reprocessing" plant. With the standard PUREX (Plutonium Uranium Extraction) reprocessing technology, the spent fuel is chopped into small pieces and dissolved in hot nitric acid. The plutonium is extracted in a light organic solvent, mixed with the nitric acid using blenders and pulse columns, and then separated with centrifugal extractors such as those that are used to separate cream from whole milk. Because all of this has to be controlled remotely from behind heavy shielding, reprocessing requires both resources and technical expertise. Detailed descriptions of the process and equipment have been published in the open technical literature since the 1950s.

Fissile Materials and Nuclear Weapons

There were nine nuclear weapon states as of 2013. In historical order they are the United States, Russia, the United Kingdom, France, China, Israel, India, Pakistan, and North Korea. The first four were part of the Cold War arms race and have been reducing their deployed arsenals from their Cold War levels. China and Israel, the fifth and sixth states respectively to make nuclear weapons, did not participate in the Cold War arms race and are believed to have kept their arsenals roughly constant for the past few decades. India and Pakistan, which carried out their first nuclear tests in 1974 and 1998, respectively, are still building up their weapon stockpiles. The status of North Korea's production complex is uncertain. The history of fissile material production in weapon states is discussed in greater detail in chapter 3. Estimates of the current stocks of nuclear warheads held by each of the nine nuclear weapon states are shown in table 2.2.

Nuclear warheads are either pure fission explosives, such as the Hiroshima and Nagasaki bombs, or two-stage thermonuclear weapons with a fission explosive as the first stage. The Hiroshima bomb was a "gun-type" device in which one subcritical piece of HEU was fired into another to make a supercritical mass (figure 2.6, left). A hollow, 16.5-centimeter (6.5-inch) outer diameter cylinder of HEU was propelled down a gun barrel to envelope a smaller cylindrical piece of HEU, creating a solid 64-kilogram cylinder of enriched uranium within a surrounding neutron-reflecting mass of natural uranium.[33] The impact of the projectile triggered a neutron source, which started an exponential fission chain reaction in the HEU that doubled every hundredth of a microsecond and, within a microsecond, released fission energy equivalent to the explosion of roughly 15,000 tons of TNT.[34]

Gun-type weapons are relatively simple devices and have been built and stockpiled without nuclear explosive tests.[35] The U.S. Department of Energy has warned that it may even be possible that intruders in a fissile materials storage facility could assemble and detonate an

Table 2.2
Estimated total nuclear warhead stocks, 2013

Country	Date of first nuclear test	Current nuclear warheads
United States	July 16, 1945	about 7,700, of which about 3,000 are awaiting dismantlement
Russia	August 29, 1949	about 8,500, with about 4,000 awaiting dismantlement
France	February 13, 1960	About 300
China	October 16, 1964	About 250
United Kingdom	October 3, 1952	About 225
Israel	September 22, 1979*	100–200
Pakistan	May 28, 1998	100–120
India	May 18, 1974	90–110
North Korea	October 9, 2006	fewer than 10

Source: Adapted from Shannon N. Kile and Hans M. Kristensen, "World Nuclear Forces," in SIPRI, *SIPRI Yearbook 2013: Armaments, Disarmament and International Security* (Oxford: Oxford University Press, 2013).
*The possible date for Israel's test is that of the "South Atlantic flash" event (see chapter 3).

"improvised nuclear explosive device" in the short time before guards could intervene.[36]

The Nagasaki bomb (and the earlier Trinity test at Alamogordo, New Mexico, in July 1945) operated using implosion. In an implosion device, chemical explosives compress a subcritical mass of material into a higher-density spherical mass. The compression reduces the spaces between the atomic nuclei and results in less leakage of neutrons out of the mass, with the result that it becomes supercritical (figure 2.6, right). Because compression reduces the critical mass, implosion-type weapons require far less fissile material than a gun-type device. They have been the preferred design for all weapon states moving beyond first-generation weapons.[37]

For either design, the maximum yield is achieved when the chain reaction is initiated at the moment when it will grow most rapidly, that is, when the fissile mass is most supercritical. This configuration is reached when the fissile components are fully assembled in the gun-type weapon or when the fissile material is most compressed in the implosion weapon.

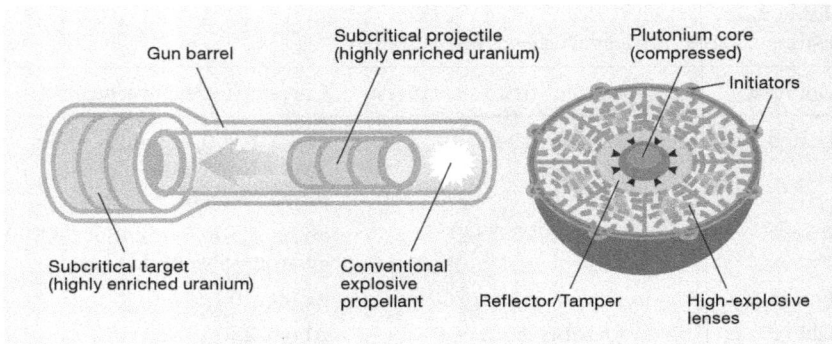

Figure 2.6

Alternative methods for creating a supercritical mass in a nuclear weapon. In the technically less sophisticated "gun-type" method used in the Hiroshima bomb (left), a subcritical projectile of HEU is propelled toward a subcritical target of HEU. This assembly process is relatively slow. For plutonium, the faster "implosion" method used in the Nagasaki bomb is required. This involves compression of a mass of fissile material. Much less material is needed for the implosion method because the critical mass decreases as the fissile material is compressed to higher density. For an increase in density by a factor of two, the critical mass is reduced to one quarter of its normal density value.
Source: Alex Wellerstein.

HEU can be used in either gun-type or implosion weapons. As is explained below, plutonium cannot be used in a gun-type device to achieve a high yield fission explosion.

Since both implosion and an envelope of neutron-reflecting material can transform a subcritical into a supercritical mass, the actual amounts of fissile material in the pits of modern implosion-type nuclear weapons are considerably smaller than bare or unreflected critical masses. Weapon state experts advising the IAEA have estimated "significant quantities" of fissile material, defined to be the amount required to make a first-generation implosion bomb of the Nagasaki type. The significant quantities are 8 kilograms for plutonium and 25 kilograms of uranum-235 contained in HEU, including losses during production of the weapon. The Nagasaki bomb contained 6 kilograms of plutonium (figure 2.7), of which about 1 kilogram fissioned in the explosion. A similar uranium-based first generation implosion weapon would contain about 20 kilograms of HEU (enriched to 90 percent uranium-235, that is, 18 kilograms of uranium-235 in HEU).

Figure 2.7

Harold Agnew on Tinian Island holding the core of the Nagasaki bomb containing 6 kilograms of plutonium.
Source: Los Alamos National Laboratory.

The United States has declassified the fact that 4 kilograms of plutonium is sufficient to make a modern nuclear explosive device. As the IAEA significant quantities indicate, an implosion fission weapon requires about three times as much fissile material if it is based on HEU rather than plutonium. This suggests a modern HEU fission weapon could contain as little as 12–15 kilograms of HEU.

Modern nuclear weapons generally contain both plutonium and HEU. The primary fission stage of a thermonuclear weapon usually contains plutonium but can contain both plutonium and HEU in a "composite" core. HEU is often used in the second stage of a thermonuclear weapon as a "spark plug" to help initiate the fusion reaction and, separately, surrounding the thermonuclear fuel to further increase the yield of the warhead (figure 2.8). Natural or depleted uranium is also used in the outer radiation case, which confines the X-rays from the primary while they compress the thermonuclear secondary.

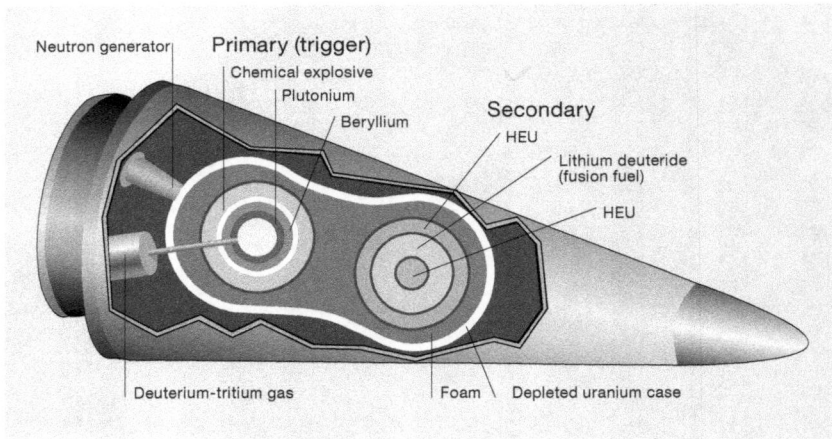

Figure 2.8

A modern thermonuclear weapon usually contains both plutonium and highly enriched uranium. Typically, these warheads have a mass of about 200–300 kilograms and a yield equivalent to hundreds of thousands of tons (hundreds of kilotons) of chemical explosive, which corresponds to about one kiloton of explosive yield per kilogram of mass. For comparison, the weapons that destroyed Hiroshima and Nagasaki weighed almost 5 tons or about 300 kilograms per kiloton of yield. The design shown here has been identified as a U.S. W-87 warhead, with a yield of 300 kilotons.
Source: Final Report of the Select Committee on U.S. National Security and Military/Commercial Concerns with the People's Republic of China, January 3, 1999.

Neutrons from thermonuclear reactions induce uranium fissions in the secondary and radiation case that can contribute one-half of the yield of the warhead.

A rough estimate of the average plutonium and HEU in deployed thermonuclear weapons can be obtained by dividing the estimated total stocks of military fissile materials possessed by Russia and the United States at the end of the Cold War by the numbers of nuclear weapons that each deployed during the 1980s: 3–4 kilograms of plutonium and 25 kilograms of HEU. In 2012, unclassified Soviet documents surfaced that indicate that the Soviet Union tested nuclear devices in 1953 containing 2 kilograms and 0.8 kilograms of super-grade plutonium with yields of 5.8 kilotons and 1.6 kilotons, respectively.[38] These were only the sixth and seventh nuclear tests that the Soviet Union had carried out at the time. The United States has declassified the fact that "all

U.S. weapon pits that contain plutonium have at least 0.5 kilograms of plutonium."[39]

In modern nuclear weapons, the yield of the fission "primary" explosive is typically "boosted" by an order of magnitude by introducing a mixed gas of deuterium and tritium, two heavy isotopes of hydrogen, into a hollow shell of fissile material (the "pit") just before it is imploded.[40] When the temperature of the fissioning material inside the pit reaches about 100 million degrees centigrade, it ignites the fusion of tritium with deuterium nuclei, which produces helium plus neutrons. This neutron burst increases the fraction of fissile materials fissioned and thereby the power of the explosion.

In a thermonuclear weapon (figure 2.8), the heat of the nuclear explosion of the primary generates X-rays that compress and ignite a "secondary" containing thermonuclear fuel, where much of the energy is created by the fusion of the light nuclei, deuterium and tritium. The tritium in the secondary is made during the explosion by neutrons splitting lithium-6 into tritium and helium.

For a time, many in the nuclear energy industry thought that the plutonium generated in power reactors could not be used for weapons because of its large percentage of plutonium-240. Plutonium-240 fissions spontaneously, emitting neutrons. This increases the probability that a neutron would initiate a chain reaction before the bomb assembly reached its maximum supercritical state. This probability increases with the percentage of plutonium-240. For gun-type designs fueled with plutonium, such "pre-detonation" would reduce the yield a thousand-fold, even for weapon-grade plutonium. The high neutron production rate from reactor-grade plutonium similarly reduces the probable yield of a first-generation implosion design—but only tenfold, because of the much shorter time for the assembly of a supercritical mass. In a Nagasaki-type design, even the earliest possible pre-initiation of the chain reaction would not reduce the yield below about 1,000 tons TNT equivalent.[41] That would still be a devastating weapon.

More modern designs are insensitive to the isotopic mix in the plutonium. As summarized in a 1997 U.S. Department of Energy report: "Virtually any combination of plutonium isotopes . . . can be used to make a nuclear weapon." The report recognizes that "not all combinations, however, are equally convenient or efficient," but concludes that "reactor-grade plutonium is weapon-usable, whether by unsophisticated proliferators or by advanced nuclear weapon states."[42] The same report continued:

At the lowest level of sophistication, a potential proliferating state or sub-national group using designs and technologies no more sophisticated than those used in first-generation nuclear weapons could build a nuclear weapon from reactor-grade plutonium that would have an assured, reliable yield of one or a few kilotons (and a probable yield significantly higher than that). At the other end of the spectrum, advanced nuclear weapon states such as the United States and Russia, using modern designs, could produce weapons from reactor-grade plutonium having reliable explosive yields, weight, and other characteristics generally comparable to those of weapons made from weapon-grade plutonium.[43]

The potential for weapons use of plutonium from nuclear power reactors makes it clear that civilian as well as military materials must be covered in any discussion of policies to reduce the dangers of fissile material.

3
The History of Fissile Material Production for Weapons

The production of highly enriched uranium and plutonium was perhaps the key challenge for the U.S. World War II nuclear weapons project. The technologies that the United States developed provided the template for the nuclear weapons programs that followed.

The United States, Russia, the United Kingdom, and France have all formally declared an end to the production of such material for weapons (table 3.1). China has let it be known that it has not produced fissile material for weapons since about 1990 but has declined to declare that this moratorium is permanent because of its concerns that the developing U.S. ballistic missile defense and offensive programs might require a buildup of China's nuclear arsenal. India, Pakistan, and North Korea continue to produce fissile material for weapon purposes, and Israel may be doing so. South Africa produced HEU for weapons, but not plutonium, and ended its HEU production when it decided to scrap its weapons. These histories are summarized in this chapter.

The fissile material production histories presented here are based on official information where available, and otherwise on unofficial information, technical articles, memoirs and news reports that have become available in the past two decades.[1] Unlike the nuclear weapon states, under the Non-Proliferation Treaty, nonnuclear weapon state parties to the treaty are obligated to declare all their nuclear materials to the International Atomic Energy Agency and make them available for IAEA monitoring. These declarations are considered confidential, however, and are not made public, except in aggregate.

United States: Setting the Precedent

The United States pursued both the HEU and plutonium routes to the nuclear bomb simultaneously because of uncertainties about each.[2] It

Table 3.1
The history of production of HEU and plutonium in countries that have made nuclear weapons

	Highly enriched uranium Production		Plutonium for weapons Production	
	Start	End	Start	End
United States	1944	1992	1944	1988
Russia	1949	1987/88	1948	1994
United Kingdom	1953	1963	1951	1995
China	1964	1987/89	1966	1991
France	1967	1996	1956	1992
Israel	?	?	1963/64	Continuing
Pakistan	1983	Continuing	1998	Continuing
India	1992	Continuing	1960	Continuing
North Korea	?	?	1986	Continuing
South Africa	1978	1990	None	None

Note: In 2012, Russia announced it was resuming HEU production to produce fuel for research reactors and fast reactors.

succeeded with both at about the same time. The bomb used on Hiroshima on August 6, 1945, was made with HEU and the bomb tested in New Mexico on July 16, 1945, and used on Nagasaki on August 9, 1945, was made with plutonium.

The United States initially explored a range of possible uranium isotope separation techniques including gas centrifuges but eventually decided to focus on electromagnetic separation and gaseous diffusion. During 1945–1947, a little over a ton of HEU was produced by gaseous diffusion feeding into electromagnetic separators at the Manhattan Project's Y-12 plant near Oak Ridge, Tennessee. U.S. HEU production shifted quickly, however, to gaseous diffusion alone. The first gaseous diffusion plant (GDP), in Oak Ridge, Tennessee, produced HEU for weapons during 1945–1964, after which it produced only low-enriched uranium for power-reactor fuel until 1985 (figure 3.1). A second GDP, in Portsmouth, Ohio, started production in 1956 and also produced HEU for weapons until 1964, when the U.S. weapons stockpile peaked at over 30,000.

Figure 3.1

Once the largest building in the world: the K-25 gaseous diffusion plant at Oak Ridge, Tennessee. The K-25 plant was the first of three gaseous diffusion plants built in the United States. Operations began in 1945 and ceased in 1985. Demolition of the plant was completed in 2013.

Source: U.S. Department of Energy.

U.S. production of HEU *for weapons* ended in 1964; new weapons could be produced from HEU being recycled from old warheads that were being retired. The U.S. enrichment complex then shifted to producing mostly low-enriched uranium for power reactor fuel and HEU enriched to over 97 percent for naval-propulsion reactor fuel.[3] HEU was produced for naval-reactor fuel through 1992 when huge quantities of weapon-grade uranium became available from excess Cold War weapons. Much of this excess HEU is now being stockpiled for future use in naval reactor fuel. Cumulatively, after subtracting HEU that was recycled through the plants, the United States produced a net of about 850 tons of HEU.

The world's first high-powered nuclear reactors were built to produce plutonium for the U.S. World War II nuclear weapons program. The first

Figure 3.2

Loading face of the Hanford B plutonium-production reactor under construction during World War II. The fuel was loaded into the ends of the tubes protruding from the graphite moderator, and irradiated fuel was pushed out of the back of the reactor into a storage pool. At a thermal power level of 250 megawatts thermal (MWt), this reactor initially produced 60–70 kilograms of weapons plutonium per year. After upgrades, it operated at more than 2,200 MWt until its final shutdown in 1968. This reactor is now a museum.
Source: U.S. Department of Energy.

weapon quantities of plutonium were produced by three graphite-moderated, water-cooled reactors on the Hanford site on the Columbia River in Washington State and were used in the nuclear test in New Mexico and the Nagasaki bomb (figure 3.2). Six additional production reactors were built later at Hanford and another five, moderated and cooled by heavy water, at the Savannah River Site in South Carolina. The primary mission of the Savannah River reactors was to produce tritium, but they also produced almost 40 percent of the U.S. stockpile of weapon-grade plutonium.[4]

As with U.S. HEU production, U.S. production of weapon-grade plutonium peaked in the 1950s and early 1960s. Nine of the fourteen U.S.

production reactors were shut down during the 1960s and early 1970s. Four of the reactors at Savannah River continued to operate into the 1980s, primarily to produce tritium for weapons. Also, one of the reactors at Hanford continued to operate but only to produce electricity. The last U.S. production reactors were finally shut down in 1988. Cumulatively, the United States produced and acquired about 110 tons of plutonium.

The United States has declared that, as of the end of September 2009, it still had about 81 tons of weapon-grade plutonium and 14 tons of non-weapon-grade military plutonium, not including material in waste. As of 2013, about 38 tons of this plutonium remains in or available for nuclear weapons, with 43 tons having been declared excess for military use. The remainder is either still in spent fuel, contained in waste, or has been disposed of.

The United States has published official histories of its production and use of plutonium and HEU. *Plutonium: The First 50 Years* describes the history of U.S. production, use, and stocks of plutonium as of the end of 1994.[5] In 2012, the plutonium report was updated through September 2009.[6] Similarly, *Highly Enriched Uranium: Striking a Balance* provides the corresponding information for HEU through the end of September 1996 and the follow-up report *Highly Enriched Uranium Inventory: Amounts of Highly Enriched Uranium in the United States* updates this information through the end of September 2004.[7]

Soviet Union/Russia: Following the Blueprint

The Soviet effort to acquire nuclear weapons began during World War II but turned into an all-out program immediately after the Hiroshima and Nagasaki bombings with the objective of breaking the U.S. monopoly on these weapons as quickly as possible.[8] The Soviet "atomic energy" effort was first made public in November 1945 and was patterned on the U.S. Manhattan Project.[9]

With regard to fissile material production, the Soviet Union drew on the official U.S. history of the Manhattan Project, *Atomic Energy for Military Purposes* (also known as the Smyth Report, because of its author Henry DeWolf Smyth, a professor of physics at Princeton University).[10] The report, issued within a week of the bombings of Hiroshima and Nagasaki, included information on the design of the plutonium production reactors[11] and the choice of the chemical process to separate plutonium from the irradiated fuel. It also discussed the different enrichment processes and barrier design considerations for the gaseous diffusion process. The report also described many problems that were encountered along the way and the general strategies used to overcome them.[12] A Russian translation of the Smyth Report was published in early 1946 and widely circulated among participants in the Soviet bomb project.[13] The Soviets decided to produce a first weapon as quickly as possible by focusing research and development on those approaches that proved successful in the Manhattan Project.

The Soviet program relied in addition on information from several participants of the U.S. project, most importantly Klaus Fuchs, who secretly passed weapon design information from Los Alamos to Moscow during the war.[14] This included detailed reports on the implosion design of the plutonium bomb and led the Soviet Union to make a virtually exact copy of that bomb for its first nuclear test.

The Soviet Union's first gaseous diffusion enrichment plant ("D-1") in Novouralsk (formerly known by a postal box code, Sverdlovsk-44) came online in November 1949.[15] As in the United States, this plant was initially complemented by a smaller plant ("SU-20") using the electromagnetic separation process for the final stage of the enrichment process to produce weapon-grade uranium. The Soviet Union eventually built three additional large enrichment plants: Seversk (Tomsk-7, 1953), Angarsk (1957), and Zelenogorsk (Krasnoyarsk-45, 1962). Initially, these plants were all based on the gaseous diffusion process and, by 1963, had a combined capacity of about 5 million separative work units per year (SWU/yr), enough to make 25 tons of weapon-grade HEU per year—enough for 1,000 nuclear weapons per year. Starting in 1964, however, the Soviet Union began introducing gas centrifuge technology and, by the early 1990s, had phased out gaseous diffusion completely (figure 3.3).

The end of HEU production in the Soviet Union probably occurred in 1987 or 1988 and was publicly announced in October 1989. By that time, the Soviet Union had accumulated the world's largest stockpile of HEU, an estimated $1,250 \pm 120$ tons of weapon-grade uranium and about 220 tons of lower-enriched material for naval, research, and fast reactor fuel.[16] Some of this material was consumed in fuel or weapons tests. By the end of 2013, a further 500 tons had been eliminated as a result of the 1993 blend-down agreement between the United States and Russia (see chapter 9). This left Russia with an estimated 665 ± 120 tons of highly enriched uranium, including material in and

Figure 3.3

Inside the Novouralsk Centrifuge Enrichment Plant. As in the United States and later the British and French plants, all Soviet plants were initially based on the gaseous diffusion process. The transition to centrifuge technology within the same buildings after 1964 permitted a sharp increase in Russia's production rate of weapon-grade uranium.
Source: U.S. Department of Energy.

available for weapons, and material reserved for naval and research reactor fuel.

The Soviet Union also surpassed the United States in plutonium production.[17] The first Soviet production-scale reactor ("A") at Mayak (Ozersk, formerly Chelyabinsk-40 and Chelyabinsk-65), modeled after the first U.S. Hanford reactors, began operations in June 1948. The Soviet Union eventually built fourteen graphite-moderated, water-cooled production reactors at three sites in Russia: six at the Mayak production complex in the Urals, five at Seversk (Tomsk-7) in Siberia, and three at Zheleznogorsk (Krasnoyarsk-26).[18] In addition, four heavy-water-moderated production reactors were operated at the Mayak site.

Russia stopped plutonium production for weapons purposes in 1994. By that time, it had produced an estimated total of 145 ± 8 tons of weapons plutonium. The stockpile remaining as of 2012 is estimated to be 128 ± 8 tons, including the 34 tons that have been declared excess for weapons purposes.[19] Russia continues to separate plutonium from power-reactor fuel at a rate of 1–2 tons per year for future use in breeder reactors.[20] Russia declared to the IAEA a stockpile of 50.7 tons of separated civilian plutonium as of December 2012.[21]

In contrast to U.S. fissile material production, which was gradually scaled back in the 1960s and 1970s, Soviet production continued at full speed almost until the Soviet Union collapsed in 1990–1991.

United Kingdom: Brothers in Arms

The U.S. and British nuclear weapon programs were intertwined from the beginning. The collaboration has included exchanges of classified weapon design information and exchanges of fissile materials for military purposes.

The history of the British weapons programs began with the Frisch-Peierls Memorandum of March 1940, the first known technical study of nuclear weapon design.[22] In this memorandum, Otto Frisch and Rudolf Peierls identified the basic steps required to make a nuclear weapon using highly enriched uranium—even though they significantly underestimated the quantity of material needed for a bomb.[23] In response to this memorandum, a British Government Committee (the MAUD Committee) recommended that work on nuclear weapons "be continued on the highest priority and on the increasing scale necessary to obtain the weapon in the shortest possible time."[24] When the scale of the undertaking became more obvious, the United Kingdom encouraged the United States to launch its own nuclear weapon program and then became an active partner in it through the participation of its scientists.

After World War II, however, the United Kingdom was cut off from the U.S. weapons program and, in 1947, started its own program. British scientists and engineers reinvented fissile material production processes and weapon designs based in part on the recollections of those scientists who had participated in the U.S. wartime effort. The postwar British weapons program was motivated primarily by the urge to maintain the status of the United Kingdom as a great power.[25]

The British effort to acquire nuclear weapons relied initially on the civilian-military ambiguity of nuclear reactors to explain the buildup of nuclear infrastructure in northwest England.[26] The construction of two air-cooled graphite-moderated reactor "piles" at the Windscale Site in

Cumbria began in 1947. The reactors became operational in 1951 and provided the plutonium for Britain's first nuclear test in October 1952. Both were shut down following a graphite fire in Pile 1 in October 1957 (figure 3.4).[27] The United Kingdom subsequently built eight dual-purpose plutonium and electric power carbon dioxide–cooled production reactors on the same site.[28] The Calder Hall and Chapelcross reactors (four units each) came online between mid-1956 and early 1960 and produced the bulk of British weapons plutonium until 1989, when the United Kingdom ended production of plutonium for weapons.[29] They also were used for tritium production.

The United Kingdom also built a dedicated military gaseous diffusion enrichment plant at its Capenhurst Site, which produced an estimated 11 ± 2 tons of highly enriched uranium from 1954 through 1962.[30] This estimate is highly uncertain because very little information is available on HEU production at Capenhurst. At the end of HEU production in 1962, the plant was reconfigured for LEU production and operated largely to enrich fuel for Britain's fleet of advanced gas-cooled power reactors until 1982, when it was replaced by a centrifuge enrichment plant.

On April 18, 1995, the United Kingdom announced that it "had ceased the production of fissile material for explosive purposes."[31] Since then, the United Kingdom has made several short declarations concerning its fissile material and nuclear weapons inventories. The 1998 Strategic Defense Review declared that the United Kingdom's military stocks of fissile materials consisted of "7.6 tonnes of plutonium [and] 21.9 tonnes of highly enriched uranium" and also declared 4.4 tons of plutonium excess to military purposes.[32] The amount of UK military plutonium therefore stands at 3.2 tons as of 2012. At an average of 4 kilograms per warhead, however, the current British nuclear weapons stockpile of approximately 200 warheads should require less than one ton of plutonium.

As of 2010, about ten tons of the United Kingdom's HEU were in submarine reactor fuel—both spent and in naval reactors. HEU not reserved for weapons is reserved for future use in submarine propulsion reactors.[33]

The 1958 US-UK Mutual Defence Agreement lifted the post–World War II wall between the British and U.S. weapons programs, formalizing an exchange of information on reactor and weapon designs and trades of British plutonium for U.S. HEU and tritium.[34] The total quantities of materials involved in these transactions have not been made public.[35] Independent analyses suggest that the United

Figure 3.4

A helicopter of the Royal Air Force taking air samples during the 1957 Windscale Fire. On October 10, 1957, the graphite core of the Windscale Pile 1 (in the background) caught fire during a routine graphite-annealing procedure, releasing substantial amounts of radioactivity including an estimated 20,000 Curies of iodine-131 and 1,000 Curies of cesium-137. Both Windscale Piles were shut down as a result of this accident. Subsequent UK production and power reactors were cooled by carbon dioxide instead of air. For estimated radioactivity releases, see Richard Wakeford, "The Windscale Reactor Accident—50 Years On," *Journal of Radiological Protection* 27 (3) (2007): 211–215; and J. A. Garland and R. Wakeford, "Atmospheric Emissions from the Windscale Accident of October 1957," *Atmospheric Environment* 41 (18) (June 2007): 3904–3920. See also "Appendix IX: Estimates of Fission Product and Other Radioactive Releases Resulting from the 1957 Fire in Windscale Pile No. 1," in Lorna Arnold, *Windscale, 1957: Anatomy of a Nuclear Accident* (New York: St. Martin's Press, 1992).
Source: Image courtesy of Louise Rawling.

Kingdom received 14–16 tons of HEU, which would be more than it produced domestically.[36]

The United Kingdom also pursued an extensive program of separating plutonium from the spent fuel of its civilian nuclear power plants—initially in the expectation that the plutonium would be used to provide initial cores for fast neutron plutonium breeder reactors.[37] However, while the United Kingdom ceased research and development efforts into fast reactors in 1994, it continued to reprocess its spent power reactor fuel. In 2009, it finally decided to end reprocessing when existing contracts had been fulfilled and began to consider how to dispose of about 100 tons of separated plutonium.[38] This is the world's largest national stockpile of civilian plutonium.[39]

France: Ambiguity by Design

France's nuclear program also built on the participation of a number of its scientists in the U.S. World War II Manhattan Project. Immediately after the war, France established an Atomic Energy Commission (*Commissariat à l'Energie Atomique*, CEA) and, in the early 1950s, began exploring the possibility of developing its own nuclear weapons program. Most significantly, in November 1954, the CEA secretly established a Nuclear Explosives Committee (*Comité des Explosifs Nucléaires*, CEN) and launched construction of plutonium production reactors at Marcoule. At the time, this was presented as a strategy to jumpstart the French plutonium breeder reactor program.[40] Production of weapon-grade plutonium began in mid-1955.

After the Suez Crisis of late 1956, during which France and the United Kingdom joined Israel in seizing the Suez Canal from Egypt, and the Soviet Union threatened them with nuclear weapons, France's government formulated a more explicit but still ambiguous position toward nuclear weapons.[41] Only when General Charles de Gaulle returned to power in June 1958—first as Prime Minister and then as President—was a weapons program fully authorized, however. Given the extensive preparations, it then proceeded rapidly. Only twenty months later, on February 13, 1960, France detonated its first nuclear weapon in the then-French Algerian desert.

France's production complex had many similarities to that of the United Kingdom: France built a series of dedicated plutonium production reactors at its Marcoule Site, three of them graphite-moderated and gas-cooled, and used several power reactors to produce additional weapon-grade plutonium and tritium. It also built a dedicated gaseous diffusion enrichment plant similar in size to the British plant at Capenhurst.

France's production of plutonium for military purposes ended in 1992[42] and, in February 1996, it announced its decision to halt production of both plutonium and HEU for weapons purposes. By the end of June 1996, the Pierrelatte enrichment plant had stopped producing HEU.[43]

France has made public very little information about its past production capacities and activities. Estimating its stocks of military fissile materials therefore remains difficult. France's current stock of military plutonium is estimated as 6 ± 1 tons and its inventory of HEU as 26 ± 6 tons.[44] Given that France has reduced its nuclear arsenal to fewer than 300 weapons, "half of the maximum number of warheads [it] had during the Cold War,"[45] and that it has shifted to LEU fuel for its naval reactors, France should have large surplus stocks of military plutonium and HEU.

France also has pursued large-scale separation and recycle of civilian plutonium in its light water power reactors.[46] Until 1997, the UP-1 plant[47] at the Marcoule Site in southern France reprocessed irradiated uranium-metal fuel from both military and civilian gas-graphite reactors.[48] A second plant (UP-2) at La Hague, on a peninsula on the northwest coast of France, started reprocessing the ceramic "oxide" fuel used in light water reactors in 1976. Later, France expanded its capacity at La Hague (UP-3 and UP2-800), financed mainly through contracts to reprocess foreign spent fuel—mostly from Germany and Japan. The licensed capacity of the La Hague plant was thus increased to 1,700 tons of spent light water reactor fuel per year, equivalent to the discharges from 80 to 90 one-gigawatt power reactors or about one-quarter of global nuclear power capacity.

France and its customers have recycled into fresh light water reactor fuel a considerable amount of the plutonium separated from their spent fuel. Nevertheless, as of the end of 2012 France had accumulated an inventory of 58.4 tons of its own separated civilian plutonium and an additional 22.2 tons belonging to foreign customers.[49]

China: The First Developing Country

China was the first country without an advanced scientific and industrial base to acquire nuclear weapons, and the first to rely on extensive, albeit short-lived, *direct* assistance from a nuclear weapon state to set up its fissile material production facilities and to train scientific personnel. China's early nuclear program was facilitated by a 1955 agreement for nuclear cooperation with the Soviet Union, which provided China with a research reactor and technical assistance and opportunities for training in Russia. A second agreement in 1958 had explicit military dimensions and included technology transfer for uranium enrichment and plutonium production.[50] In 1958, China started the construction of the Lanzhou gaseous diffusion plant.

A December 1960 U.S. National Intelligence Estimate on the status of China's nuclear program and capabilities observed that China had a "shortage of trained scientists and engineers." It listed only ten nuclear physicists working in China at the time and concluded that "this shortage would hamper Chinese efforts to design, construct and operate facilities for the production of fissionable materials and would be particularly serious, should the Soviets decide to reduce or terminate their technical aid."[51] In fact, the Soviet Union already had withdrawn its technical experts in August of that year.[52]

The official history of China's nuclear program reports that "when the Soviet Union scrapped its treaties . . . the primary link in the U-235 production line—the Lanzhou Uranium Enrichment Plant—had been basically completed and the equipment complement was rather complete," but Chinese scientists and engineers found completing the uranium enrichment plant "extremely complex and difficult."[53] To make matters more challenging, "the primary technical equipment for the diffusion plant had not been fully outfitted, and some important technical data had been carried back to the USSR by the experts or burned."[54] The Soviet Union also was supposed to have provided uranium hexafluoride feed material for the enrichment plant but China had to learn how to produce this for itself. The Lanzhou plant finally began to produce weapon-grade HEU in early 1964, and China's first uranium-based nuclear explosion was carried out later that year. The Lanzhou plant ended HEU production in 1980.

China's second gaseous diffusion plant, the Heping facility, is believed to have started operating around 1975 and may have had an initial capacity about the same as the Lanzhou plant, but the capacity was increased over time. The Heping plant ended HEU production in about 1987.

China produced an estimated 20 ± 4 tons of HEU, a few tons of which were consumed in nuclear tests and reactor fuels. Its current stockpile is estimated at about 16 ± 4 tons of HEU.

Like other countries, China has moved from gaseous diffusion to gas centrifuges to produce low-enriched uranium for its power reactors. As of 2013, China had three Russia-supplied centrifuge plants, with a combined capacity of 1.5 million SWU/yr.[55] In 2010, China also began operating an enrichment plant with indigenous centrifuges, with a capacity in 2013 estimated at 1 million SWU/yr.[56]

Starting in 1958, the Soviet Union also helped China with its plutonium production program, supplying assistance both with a production reactor and a reprocessing plant. The Jiuquan plutonium production reactor is a graphite-moderated, water-cooled reactor, probably modeled on the Soviet Mayak reactors. Construction of the reactor started in 1960 but was delayed when Soviet assistance ended that year. It was completed in 1966.

The Soviet Union provided China with a design for a reprocessing plant, but the design apparently did not feature the PUREX method, which the United States had developed in the early 1950s and which had quickly became the standard worldwide. China's nuclear leadership was dissatisfied with the Soviet design and, in 1959, launched its own parallel effort to master reprocessing.[57] China began to build a military reprocessing plant at the Jiuquan site in 1964, which was completed in April 1970 but only started operation in 1976 due to "certain shortcomings in the designs."[58]

In 1967, China started construction on three 80-megawatts thermal (MWt), graphite-moderated, water-cooled plutonium-production reactors in mined caverns under a mountain near Fuling to provide protection from a possible U.S. or Soviet attack. But progress was slow and eventually the project was abandoned. Instead, a copy of the Jiuquan reactor was built near Guangyuan. The new reactor went critical in December 1973. In 2010, the partially completed underground reactor complex at Fuling was opened to the public as a tourist attraction.[59]

China produced an estimated total of 2.0 ± 0.5 tons of weapon-grade plutonium.[60] An estimated 200 kilograms of this plutonium were consumed in China's nuclear tests, leaving a stockpile of 1.8 ± 0.5 tons of plutonium for weapons.[61]

China has indicated informally that it has suspended all production of fissile material for weapons.[62] But it appears to have a policy for possibly resuming such production if required.[63]

China also has started a civilian program to reprocess spent power reactor fuel and use the separated plutonium to fuel breeder reactors for electricity generation. A civilian pilot reprocessing plant was commissioned in Gansu Province in December 2010 with a capacity of 50–60 tons of spent power reactor fuel per year.[64] A stock of 13.8 kilograms of plutonium had been separated at this plant as of December 31, 2012.[65] China has been discussing for several years plans for either purchasing from France or building a domestically designed commercial-scale civilian reprocessing plant but the cost would be high and a government decision has been delayed.

Israel: A Turnkey Nuclear Weapons Program

Israel launched its nuclear weapons program at its Dimona site in the 1950s, with comprehensive secret assistance from France. It was the first time that a nuclear weapon state supplied a complete fissile material production complex, including a turnkey nuclear reactor and reprocessing plant, for the explicit purpose of jump-starting a nuclear weapons program of an ally.[66] France reportedly also shared information with Israel on the design and manufacture of nuclear weapons.[67] To this date, however, Israel does not acknowledge its possession of nuclear weapons. In addition to plutonium, the Dimona reactor is believed to produce tritium and the site may also host a small gas centrifuge uranium enrichment plant.[68]

The only insider revelations about operations at Israel's nuclear facility at Dimona were published in a front-page article in the London *Sunday Times* in October 1986, based on information supplied by Mordechai Vanunu, who had been employed as a technician there from 1977 until 1985. Vanunu was then kidnapped in Italy, tried in Israel for treason, and spent eighteen years in prison. Vanunu's notes and his collection of about sixty pictures taken inside the Dimona complex (figure 3.5) are the basis for the assessment of Israel's plutonium production summarized below. Experts have judged the technical details in the unpublished transcripts of the interviews with Vanunu to be credible. Some of the information, however, is difficult to make consistent.[69]

Estimating Israel's plutonium production and its current stockpile of fissile material hinges on the history of the power level of the Dimona plutonium production reactor. This heavy water "research reactor" was originally reported to have a design power of 24–26 MWt. There is evidence for power uprates early on, however, and it is widely assumed that the reactor has operated at 70 MWt or more since the 1970s.[70] Vanunu's interview suggests a net plutonium production rate of 36 kilograms per year, which would correspond to a power level of about 140 MWt.

Figure 3.5

Two of the pictures taken by Vanunu inside Dimona in or before 1985, showing mock-up bomb components. Vanunu shared these photos, along with his notes about the operation of the facility, with reporters for the London *Sunday Times*. A front-page story, based on this information, was published by the *Times* on October 5, 1986. By that time, however, Vanunu had been kidnapped by Israeli intelligence agents and taken to Israel where he was tried in secret and sentenced to eighteen years in prison.

Source: Authors' archives.

Overall, as of the end of 2012, cumulative plutonium production at Dimona is estimated to be 840 kilograms ± 125 kilograms.[71] This would be sufficient for about 160 weapons, which may be enough to satisfy Israel's perceived national security needs. Plutonium production at Dimona may be continuing as of 2013 largely as a "by-product" of production of tritium for use in weapons. Unlike plutonium, tritium has to be replenished because it has a half-life of twelve years.

Israel appears to have relied on outside sources of uranium to support the operation of the Dimona reactor. In 1968, Israel covertly acquired about 200 tons of uranium ore from Belgium by transferring it at sea from one ship to another in what became known as the Plumbat affair.[72] More significantly, starting in 1965, South Africa provided Israel 500 tons of uranium. This material remained under bilateral safeguards until 1976 when South Africa agreed to lift these safeguards and to deliver 100 additional tons of uranium oxide yellowcake (equivalent to about 70 tons of uranium) in return for 30 grams of tritium.[73] Thus South African assistance in supplying uranium was critical to keeping Israel's Dimona reactor operating for many years. The total amount of South African supplied uranium was sufficient to produce about 400 kilograms of weapon-grade plutonium, about 50 percent of Israel's estimated plutonium stockpile.

It appears also that Israel clandestinely acquired up to a few hundred kilograms of weapon-grade HEU from a naval fuel fabrication plant in the United States.[74] Israel's interest in uranium enrichment also is well documented.[75] There is no published evidence, however, that the enrichment programs described by Vanunu as based on centrifuge and laser technologies went beyond research and development.

India: The Cover of the Peaceful Atom

India's nuclear program was launched soon after it became independent in 1947 and in the shadow of the atomic bombings of Hiroshima and Nagasaki. From the outset, the nuclear program was ambiguous about its purposes, claiming to be aimed at producing electricity but making technology choices aimed at keeping open the weapons option.

India chose the plutonium path because, by the mid-1950s, international assistance was available through Atoms for Peace for acquiring reactors and reprocessing technology but not enrichment.[76] India's first plutonium production reactor was the 40 MWt, heavy-water-moderated, light-water-cooled, natural-uranium-fueled *CIRUS* reactor provided by

Canada and based on the Canadian *NRX* reactor.[77] The heavy water was supplied by the United States. *CIRUS* went critical in 1960 and began operating at full power in 1963.

Under the terms covering the supply of the reactor, India committed to use it and the fissile material coming out of it for peaceful purposes,[78] but later used plutonium separated from spent CIRUS fuel for its 1974 test nuclear explosion. Despite India's description of the test as a "peaceful nuclear explosion," Canada and the United States suspended nuclear cooperation with India for three decades. Cooperation with the United States was only restored in 2008 as part of the U.S.-India nuclear deal, which drove subsequent controversial changes to the guidelines of the 44-nation Nuclear Suppliers Group, lifting international restrictions on nuclear trade with India. As part of this deal, in December 2010, the CIRUS reactor was shut down.

In 1985, a second plutonium production and research reactor entered service next to *CIRUS*. *Dhruva* has a thermal power of 100 MWt and is heavy water cooled and moderated. India is planning to build another reactor with similar power to enter service in 2017–2018.

India also copied and later adapted the design of the two 220-megawatts electrical (MWe) pressurized heavy water reactors (PHWRs) provided by Canada in the 1960s to build a fleet of sixteen additional heavy water power reactors as of the end of 2013.[79] These include eleven 220 MWe and two 540 MWe PHWRs that are unsafeguarded. The first discharges of spent fuel from such reactors have low burnup and contain plutonium with over 90 percent plutonium-239; that is, it is weapon-grade.[80]

It is estimated that India has produced a total of 620 ± 180 kilograms of weapon-grade plutonium. Some of this plutonium was used in the nuclear tests of 1974 and 2008 and also for the first core of the India's plutonium-fueled Fast Breeder Test Reactor (FBTR), which was commissioned in 1985. As of 2012, India was estimated to have a stockpile of 540 ± 180 kilograms of weapons plutonium.

India's fleet of unsafeguarded PHWRs also had given it a stockpile (as of the end of 2012) of an estimated 4.7 ± 0.4 tons of separated power reactor plutonium, almost ten times more than was in its stockpile of weapons plutonium from dedicated production.[81] India states that this plutonium is intended to fuel a fleet of breeder reactors for electricity production. India's 500-MWe Prototype Fast Breeder Reactor (PFBR) is outside IAEA safeguards, however. It is to be fueled by reactor-grade plutonium and could produce more than 100 kilograms of unsafeguarded weapon-grade plutonium per year by neutron capture in a natural uranium "blanket" surrounding the core.[82] India has plans for four more breeder reactors by 2020.

The design of India's oldest reprocessing plant at the Bhabha Atomic Research Centre in Trombay was supplied by the United States for peaceful purposes but has been used for separating plutonium for weapons. India built two additional reprocessing plants in 1977 and 1998 mostly to supply startup plutonium for breeder reactors.[83] Another reprocessing plant was added in 2010.[84] An additional "fairly large" reprocessing plant had been under construction for several years as of 2013.[85] India's Department of Atomic Energy expects to build over the next decade up to three more reprocessing plants with capacities close to 500 tons per year each.[86] It appears unlikely for the foreseeable future that these plants will be safeguarded.

Although it initially considered uranium enrichment as too demanding, India in the 1970s began to develop centrifuge enrichment technology. India now has two gas centrifuge uranium enrichment facilities: a pilot scale plant that has been operating since 1985 and a larger production scale plant at Rattehalli near Mysore that has been operating since 1990. Rattehalli is believed to produce HEU enriched to 30–45 percent uranium-235 to fuel India's nuclear-powered submarines. As of 2012, India's estimated HEU stockpile was 2.4 ± 0.8 tons.

India has developed new generations of more powerful centrifuges and its centrifuge production capacity has been increased. It is planning a second enrichment plant, the "Special Material Enrichment Facility," in Chitradurga district in Karnataka. According to the chairman of India's Atomic Energy Commission, this facility will not be offered for IAEA safeguards since India is "keeping the option open of using it for multiple roles."[87] These roles may include producing HEU for nuclear submarines and for weapons in addition to low-enriched uranium for imported civilian power reactors.

Pakistan: Turning to the Nuclear Black Market

Pakistan launched its nuclear weapon program in January 1972, with plans to pursue both uranium enrichment and plutonium separation. The program gained more urgency after India's first nuclear test in May 1974. Progress was limited however by the fact that Pakistan had a very low technology base. China has provided material and technology support. Pakistan also has relied heavily on clandestine purchases from commercial suppliers in other countries of key technologies and components for its fissile material production program.

In 1974, Pakistan began work on a gas centrifuge enrichment program.[88] Progress was made possible by the illicit acquisition of centrifuge design information, components, and expert assistance organized by Pakistani metallurgist A. Q. Khan. Khan's employment at a supplier for the Dutch branch of the URENCO uranium enrichment company gave him access to technical information on Dutch and German centrifuge designs.[89]

In 1982, Pakistan reportedly enriched uranium up to weapon-grade.[90] Its success was made possible by extensive purchases in the 1970s of the key components and materials for its centrifuge program from commercial suppliers in Germany, Switzerland, the Netherlands, the United Kingdom, France, the United States—and perhaps other countries as well.[91] These included vacuum pumps and valves, equipment for uranium hexafluoride production (including conversion and purification, gasification and solidification), high-strength aluminum and steel for centrifuge rotors and components, and high-frequency inverters for centrifuge motors. Pakistan also relied on European centrifuge experts who were willing to consult on technical problems. In addition, A.Q. Khan claims that Pakistan received 50 kilograms of weapon-grade HEU and a nuclear weapon design from China.[92]

There is great uncertainty in the estimate of Pakistan's HEU stocks because of the lack of reliable information about both the operating history and enrichment capacity of Pakistan's centrifuge plants. A midrange estimate is that, as of the end of 2012, Pakistan had produced about 3,000 kilograms of HEU and had consumed up to 100 kilograms of this in its 1998 nuclear weapon tests.

Pakistan also looked abroad to acquire facilities, technologies, and equipment for plutonium separation.[93] In 1967, it obtained from the UK Atomic Energy Authority the design for an experimental reprocessing facility to be located at the Pakistan Institute of Nuclear Science and Technology (PINSTECH), Islamabad, which already housed a 5-megawatt HEU-fueled research reactor provided in 1963 by the United States under the Atoms for Peace program.[94]

After the experimental facility was completed in 1971, Pakistan sought a larger pilot reprocessing plant. When the United Kingdom refused, Pakistan turned to the Belgian company Belgonucléaire and then to the French company Saint-Gobain Techniques Nouvelles (SGN), which together supplied a pilot reprocessing plant that was sited close to

PINSTECH and became known as the "New Labs."[95] The plant was completed in 1982 but may not have been operated for about two decades, since Pakistan did not yet have an unsafeguarded plutonium production reactor. A second reprocessing plant at the New Labs, which looks similar in size to the original plant, was completed in 2006.[96]

Pakistan signed a new contract with SGN in 1974 for a much larger 100 ton per year reprocessing plant. In 1978, however, under U.S. pressure, France canceled the contract. Significant design information and some technology had been transferred and some infrastructure construction undertaken before the cancellation, however.[97] Work apparently did not resume at the site until after 2000–2002, but there is no evidence the plant is operational.[98]

Work on Pakistan's first plutonium production reactor, Khushab I, started in the early 1980s and it came online in 1998.[99] It is a heavy-water-moderated natural-uranium-fueled reactor with a thermal power of about 40–50 MWt[100] capable of producing about 10 kilograms of plutonium a year. China is believed to have helped in its construction.[101] Pakistan has been replicating the Khushab reactor. As of 2013, Khushab I and II were operating, Khushab III was approaching completion, and Khushab IV was under construction.

Spent fuel from Khushab I and II is reprocessed at the two New Labs reprocessing facilities near Rawalpindi, which together have an estimated capacity to reprocess 20–40 tons per year of irradiated uranium.[102] This corresponds to a rate of separation of 16–32 kilograms of weapon-grade plutonium annually. In early 2000, air samples, reportedly collected by the United States, showed traces of krypton-85, indicative of active reprocessing.[103]

As of the end of 2012, Pakistan could have produced cumulatively about 150 ± 50 kilograms of plutonium for weapons. Pakistan's annual rate of production of plutonium for weapons will likely at least double when the third and fourth Khushab production reactors come online. Pakistan has no civilian plutonium separation program as yet.

North Korea: A Nuclear Bargaining Chip?

North Korea began to develop its nuclear program in the wake of the Korean War (1950–1953), during which the United States publicly threatened the use of nuclear weapons.[104] Like China, North Korea sent scientists and engineers to the Soviet Union for training and acquired a research reactor from the USSR. It built a small plutonium production reactor at Yongbyon, which began operating in 1986. The design of the 25 MWt carbon-dioxide gas-cooled graphite-moderated natural-uranium fueled reactor appears to have been based on the United Kingdom's "Magnox" reactors whose design was made public in the 1950s.

The North Korean reactor operated from 1986 to 1994 and then was shut down under the Agreed Framework deal with the United States that would roll back North Korea's nuclear program in exchange for two light water power reactors. As part of this deal, North Korea suspended operation of the Yongbyon reactor, stopped construction on the associated reprocessing plant, and halted work on two much larger plutonium production reactors. The deal collapsed in 2002 after the United States accused North Korea of having a secret HEU production program and, in January 2003, North Korea withdrew from the NPT—the first country to do so—announced that it was developing nuclear weapons, resumed operation of the Yongbyon reactor, and began to reprocess its previously discharged spent fuel.

In what became known as the Six-Party Talks, North Korea negotiated with the United States, South Korea, Japan, China, and Russia on steps to reverse its nuclear weapon program. This led in 2007 to agreement on a plan that included early disablement of key nuclear facilities, including the Yongbyon reactor. As part of this plan, the cooling tower for the reactor was demolished. Also as part of the plan, North Korea provided to China in June 2008 a declaration of how much plutonium North Korea had separated—variously reported in the press as 31 kilograms and 37 kilograms—and provided operating records.[105] This is roughly consistent with independent estimates of plutonium production in North Korea.[106] Negotiations subsequently broke down, however, on a verification agreement in which the United States proposed to take graphite samples from the reactor core to check the accuracy of the declaration.[107]

The agreement on disablement broke down in 2009 and North Korea announced that it would reprocess its remaining spent fuel at Yongbyon. North Korea separated an additional 10 kilograms or so of plutonium from this fuel.[108] This suggests that as of 2012 North Korea may have had a stock of about 30–40 kilograms of separated weapon-grade plutonium.[109] North Korea appears to have restarted the Yongbyon reactor in late 2013.

North Korea also has pursued centrifuge uranium enrichment with assistance from Pakistan.[110] According to a November 2002 declassified U.S. Central Intelligence Agency assessment, "North Korea was constructing a plant that could produce enough weapon-grade uranium for two or more nuclear weapons per year when fully operational—which could be as soon as mid-decade."[111] In 2010, North Korea revealed a uranium enrichment plant at the Yongbyon site, with an estimated 2000 centrifuges.[112] The plant had been built since an IAEA inspection in April 2009. North Korea stated that the enrichment plant is intended to produce LEU for a light water reactor that is being built on the same site. North Korea may, however, have in addition a covert enrichment plant or plants that could be used to produce HEU for weapons.

South Africa: Coming in from the Cold

In the early 1960s, South Africa's white minority government developed a nuclear enrichment capability based on a novel but inefficient enrichment technology. During the 1980s, under threat from Cuban army units flown into neighboring Angola by the Soviet Union, South Africa built an arsenal of seven HEU nuclear weapons, based on a simple gun-type design. In the early 1990s, while preparing to transfer power to a majority government, South Africa dismantled its weapons and related production facilities and joined the NPT as a non-weapon state. It is the only example thus far of a state that developed nuclear weapons and then, as political circumstances changed, voluntarily disarmed.[113]

South Africa's nuclear history started in the mid-1940s as part of the British and U.S. effort to acquire uranium for the U.S. nuclear weapons program. Substantial low-grade uranium was soon discovered in existing gold mines. South Africa began a nuclear research and development program in 1961 at the Pelindaba Site, near Pretoria. As part of its Atoms for Peace program, the United States supported construction of the Safari-I research reactor and provided it with HEU fuel. Safari-I was commissioned in 1965 and remains operational as of 2013, but has been converted to LEU fuel.[114]

South Africa explored both plutonium production and uranium enrichment and, in 1969, launched a uranium enrichment program.[115] The program used an aerodynamic separation process known as the Helikon aerodynamic vortex tube or "stationary-walled centrifuge."[116] This method was used at the Pelindaba Site in a pilot plant to produce highly enriched uranium (Y-Plant) and later, in a much larger "semi-commercial" plant at the same site (Z-Plant), to produce low-enriched uranium for a commercial power reactor.[117]

Construction of the Y-Plant, which would later support South Africa's weapons program, began in 1971, and the first HEU, enriched to 80 percent uranium-235, became available in January 1978.[118] Justification for the enrichment program included production of 45 percent enriched uranium for the Safari-I research reactor after U.S. fuel supplies were terminated.[119] There were suspicions about a possible military dimension of the program, however, because South Africa refused to join the NPT.

South African officials involved in the project have stated that the initial objective of the program was to develop the technologies for "peaceful nuclear explosions," which were being promoted in the 1960 and 1970s by the U.S. and the Soviet nuclear-weapon laboratories. In 1977, South Africa began to prepare a test shaft for a "cold nuclear test" in the Kalahari Desert.[120] These test preparations were detected by the Soviet Union,[121] which informed the United States. Under diplomatic pressure, South Africa abandoned this effort.

South Africa's first complete nuclear weapon, a gun-type highly enriched uranium design with an estimated yield of about 10 kilotons, was ready in December 1982, and about one warhead was added to the arsenal each year thereafter.[122] In 1985, President P. W. Botha ordered that South Africa's arsenal be limited to a total of seven devices. Scientists in the weapon program continued limited research on more advanced weapons including implosion and boosted designs.

In 1990, shortly after assuming office, President F. W. de Klerk ordered the end of the weapons program.[123] After dismantlement of the weapons, recovery of the HEU, and destruction of sensitive documents, South Africa joined the NPT in 1991 and signed a comprehensive safeguards protocol with the IAEA later that year. Only in March 1993 did President de Klerk make public that South Africa had had nuclear weapons.[124]

The HEU produced by South Africa between 1978 and 1990 was placed under IAEA safeguards, but the amount has not been made public. At a minimum, a stockpile of about 400–450 kilograms of HEU must have been available, in order to make the cores for the seven gun-type devices in South Africa's arsenal.[125] Other estimates have been significantly higher, up to 680–790 kilograms.[126] Since then, a significant fraction of this HEU stockpile has been consumed in civilian applications.[127]

It took the IAEA several years to satisfy itself that "there were no indications to suggest that the initial inventory is incomplete or that the nuclear weapon programme was not completely terminated and dismantled."[128] The South African case provides a precedent for the international verification of nuclear disarmament, but the task would be much more difficult for countries originally possessing hundreds or thousands of nuclear weapons.

4

The Global Stockpile of Fissile Material

Seven decades of effort to produce fissile materials in large amounts for both military and civilian programs have left an enormous global stockpile of highly enriched uranium and plutonium. As of the end of 2012, this combined stockpile was about 1,900 tons, about three-quarters of which was HEU and the rest plutonium.

As of 2013, almost all of the HEU stockpile and about half of the plutonium stockpile were produced for weapon purposes. Some of the weapon states produced additional HEU for naval and research reactor fuel. Since the 1970s, some non-weapon states also acquired the capability to separate plutonium and enrich uranium as part of their civilian nuclear power programs. This chapter provides an overview of the global stockpiles of HEU and separated plutonium, focusing on the amounts of material in different categories as determined by current or intended use. National stockpiles of highly enriched uranium and of plutonium as of the end of 2012 are shown in the appendix to this chapter.

As described in chapter 3, the history of this production is in some cases not well understood because of the secrecy surrounding military nuclear programs. This has led to significant uncertainties about the size of some national stockpiles. Increased transparency by all weapon states will be required to provide essential background information for the negotiation and verification of deep reductions in the nuclear arsenals and the eventual elimination of nuclear weapons.

In 2006, the UK government noted: "Transparency about fissile material acquisition for defence purposes will be necessary if nuclear disarmament is to be achieved; since achieving that goal will depend on building confidence that any figures declared for defence stockpiles of fissile material are consistent with past acquisition and use."[1]

The United States offered a different but also important set of reasons in 1994 to justify the official publication of its history of HEU production and use, stating:

The American public will have information that is important to the current debate over proper management and ultimate disposition of uranium . . . The quantities may aid in public discussions of issues related to uranium storage safety and security. The data will be of some aid to regulators who will oversee environmental, health and safety conditions at the national laboratories [and] have valuable nonproliferation benefits by making potential International Atomic Energy Agency safeguards easier to implement.[2]

Independent analysts also have emphasized the importance of nuclear transparency and the role it can play in various arms control contexts, including in reducing the discriminatory nature of the nonproliferation regime under which only NPT non-weapon states are legally required to report their nuclear material holdings and allow verification by the International Atomic Energy Agency.[3]

These concerns were recognized in the "Action Plan on Nuclear Disarmament" agreed at the 2010 Non-Proliferation Treaty Review Conference, which affirmed that "nuclear disarmament and achieving the peace and security of a world without nuclear weapons will require openness and cooperation, and . . . enhanced confidence through increased transparency and effective verification."[4]

Taken together, all these reasons make a compelling case for nuclear weapon states to become more transparent about their national fissile material stockpiles.

The 1998 Guidelines for the Management of Plutonium established one limited transparency agreement. Under these guidelines (published by the IAEA as INFCIRC/549[5]), the United States and United Kingdom, China, France, and Russia along with four non-weapon states (Belgium, Germany, Japan, and Switzerland) make annual public reports of their civilian plutonium stockpiles to the International Atomic Energy Agency. Since 2001, France and the United Kingdom also have declared their civilian HEU holdings.

Highly Enriched Uranium

The global stockpile of highly enriched uranium was about $1,380 \pm 125$ tons at the end of 2012, enough for more than 50,000 simple, first-generation implosion weapons. About 98 percent of this material is held by the nuclear weapon states, with Russia and the United States having

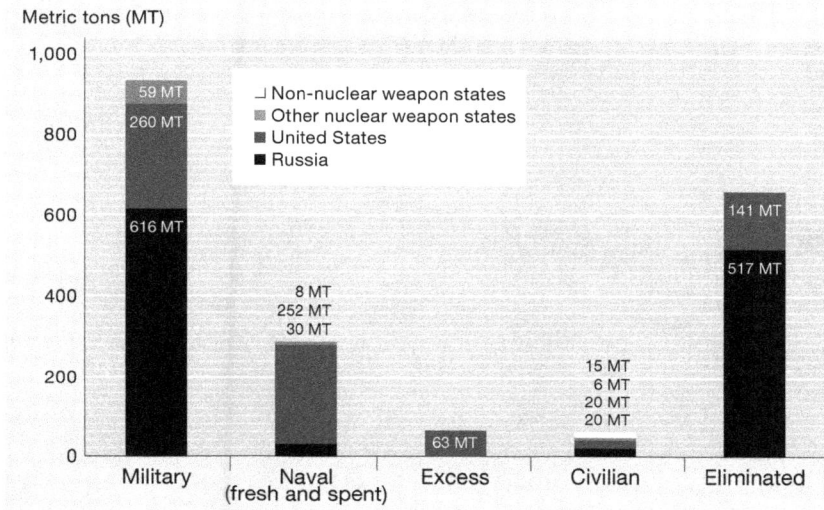

Figure 4.1
Global stockpiles of highly enriched uranium by category as of the end of 2012. Military stocks of HEU are IPFM estimates with the exception of the inventories of the United States and the United Kingdom, which are based on their public declarations. The uncertainty in the global military stockpile of HEU is on the order of ±100 tons. The different stockpile categories shown in the figure are discussed in the text. Russia's stockpile has been updated to the end of 2013, when the agreement with the United States to blend down 500 tons of excess weapon-grade uranium was completed.

by far the largest HEU stockpiles (figure 4.1 and see the appendix to this chapter). As a consequence, almost all HEU remains outside IAEA safeguards. In 2013, only India, Pakistan, Russia, and possibly North Korea were believed to be producing HEU. These production programs were relatively small scale compared to the Soviet and U.S. Cold War programs, however, and, since Russia and the United States had together blended down about 660 tons of excess HEU since the end of the Cold War, the global HEU stockpile was down considerably from its Cold War peak.

HEU Available for Weapons
The estimated HEU inventory available for weapons was 935 ± 100 tons. There was ample scope for additional HEU to be declared as excess for weapons purposes. Approximately 17,000 nuclear warheads existed in

Figure 4.2
The U.S. HEU Material Facility (HEUMF) at Y-12, Oak Ridge, Tennessee. The facility provides storage capacity for 12,000 drums and 12,000 cans of material in specially designed storage racks and can hold up to 400 metric tons of highly enriched uranium.
Source: B&W Y-12.

the nuclear weapons arsenals. If, on average, each warhead contained 25 kilograms of HEU, then about 500 tons were in nuclear weapons and reserve components worldwide. This left over 400 tons of HEU in the weapons complexes without any apparent use. This material could be declared excess for weapons purposes. The United States stores uranium components (secondaries) in large numbers at a dedicated facility at the Y-12 site in Oak Ridge, Tennessee (figure 4.2).

Naval HEU
The five NPT weapon states and India operate nuclear-powered military ships and submarines. The United States, Russia, the United Kingdom, and India use highly enriched fuel for this purpose (see chapter 7 for a detailed discussion). The United States and the United Kingdom use weapon-grade uranium for naval fuel. Russia uses a variety of HEU enrichment levels, but the fuel is typically not weapon-grade. India reportedly uses 30–45 percent-enriched fuel. France has moved to

low-enriched uranium submarine fuel, and China is also believed to use LEU fuel for its nuclear fleet.

Naval use of HEU has led to a significant stockpile of material in this category, about 180 tons. The United States has earmarked for future use in naval fuel nearly its entire stockpile of weapon-grade uranium declared excess for weapons purposes and that meets navy specifications. In 2005, for example, the United States declared an additional 200 tons of HEU excess for weapons purposes, but it reserved 152 tons of that material for future use in naval fuel. Similarly, the United Kingdom has stated that HEU that is no longer needed for weapons purposes will be reserved for naval fuel.[6] Russia too uses HEU to fuel its fleet of naval vessels, but it does not formally identify the stocks reserved for this purpose and may only assign material on demand.

In principle, naval stockpiles of HEU could be offered for IAEA safe-guards—at least until they are fabricated into fuel. Following the HEU into the fuel fabrication process and then to the reactors would be complicated by the fact that the designs of naval fuel and reactors are often considered sensitive. Moreover, part of the material assigned to naval reserves may still be in classified form associated with its former use as nuclear weapon components. These complications also will arise in the context of verifying a future Fissile Material Cutoff Treaty and are discussed in chapter 8.

The most direct approach to minimizing the sizes of naval HEU reserves would be to end the use of HEU for naval fuel altogether. This is discussed in more detail in chapter 7. The next generations of U.S., Russian, and British naval vessels are already designed to be HEU fueled, however, and U.S. and UK vessels will be equipped with lifetime cores.

The use of nuclear naval propulsion could spread over the next decades. Brazil is the first non-weapon state that is actually developing a reactor for submarines.[7] Such a trend underscores the importance of having high-performance LEU fuels available for this application.

Spent Naval HEU Fuel

The uranium in spent naval fuel that was initially weapon-grade is still highly enriched in uranium-235 and so continues to be HEU, making it a significant contributor to the global stockpile.[8] The United States and Russia have used the largest amounts of HEU fuel for their fleets of nuclear submarines and aircraft carriers. Russia reprocesses most of its spent naval fuel to recover the residual HEU for further use, however, and only a small fraction (on the order of 10 tons) remains in extended storage as of 2013.[9] The United States stopped reprocessing its naval reactor fuel in 1992. As of 2012, there were about 30 tons of highly enriched uranium stored in spent naval reactor fuel at Idaho National Laboratory.[10] The current plan is to dispose of this material in a geological repository without further processing. The United Kingdom also stores its spent naval fuel. Stockpiles in this category will continue to increase and strategies will have to be found to store and manage them safely and securely, ideally under international monitoring.

Civilian HEU
The global stockpile of civilian HEU stands at about 60 tons, roughly equally distributed between the United States, Russia, and all other countries taken together. Estimates are difficult to make because the distinction between civilian and non-civilian material in weapon states is not always clear-cut, while non-weapon state reports of their HEU inventories to the IAEA are considered confidential information and not published.[11]

International efforts to phase out the use of HEU in the civilian nuclear fuel cycle have been under way since the late 1970s following establishment of the Reduced Enrichment for Research and Test Reactor (RERTR) program and have intensified since 2001.[12] They are discussed in greater detail in chapter 7. These activities, now spearheaded by the U.S. Global Threat Reduction Initiative, have focused on converting research reactors to low-enriched fuel and repatriating fresh and spent HEU fuel to Russia and the United States. The spent fuel is then stored awaiting final disposal or reprocessed with the recovered HEU downblended to LEU. In many cases, reactor conversion has been delayed for both technical and political reasons. Given the emphasis that has been given to the issue,[13] the use of HEU in the civilian nuclear fuel cycle could be largely phased out by 2025.

Excess HEU
Since the 1990s, some weapon states have declared significant amounts of HEU excess for weapons or all military purposes. Most notably this includes the landmark 1993 agreement between Russia and the United States to blend down 500 tons of excess Russian weapon-grade HEU to LEU by the end of 2013.[14] For its part, in 1994, the United States declared excess to all military purposes 174 tons of HEU and, in 2005 declared an additional 200 tons of HEU excess to weapons purposes. Virtually all material in the U.S. declarations that was weapon-grade—namely, that contained more than 90 percent uranium-235—was earmarked for future

use in naval fuel. There are currently no plans for any transparency measures on this naval stockpile. U.S. excess HEU that does not meet navy specifications is being blended down or awaits direct disposal.

As of the end of 2013, Russia had down-blended all 500 tons that were part of the original U.S.-Russian Agreement.[15] The remaining HEU committed for blend down (about 60 tons) was U.S. material, which may not be blended down entirely for decades.[16]

Eliminated HEU

As of the end of 2013, about 660 tons of HEU had been eliminated. This corresponds to about one third of the HEU stockpile that existed at the end of the Cold War. The additional HEU that had been declared excess brought the amount of HEU eliminated or to be eliminated up to 720 tons. In principle, the global HEU stockpile could be reduced much further, especially if naval propulsion moved away from HEU fuel. For example, only about 100 tons of HEU would be required to support a global stockpile of 4,000 nuclear warheads. If the nuclear weapon states reduced arsenals to that level, more than 90 percent of the global stock of HEU as of 2013 could eventually be eliminated. The economic value of HEU when blended down to low-enriched uranium for use in power reactor fuel provides an incentive to dispose of excess stocks instead of maintaining them as "fissile material hedges" that they will most likely never need.

Separated Plutonium

The global stockpile of separated plutonium in 2012 was about 492 ± 10 tons. About half of this stockpile was produced for weapons, while the other half was separated from spent power reactor fuel in civilian reprocessing programs (figure 4.3). Only India, Pakistan, and perhaps Israel and North Korea continue to produce plutonium for weapons purposes. Very little progress with regard to plutonium disposition has been made since the end of the Cold War. Disposition of Russian and U.S. excess weapons plutonium via irradiation in reactor fuel was planned to begin by 2007 but has been delayed.

Plutonium Available for Weapons

As of 2012, almost 140 tons of plutonium remained available for weapons. For a global inventory of 10,000 warheads (not including warheads in the dismantlement queues and stored plutonium weapon components or "pits") and an average of 5 kilograms per warhead, which is

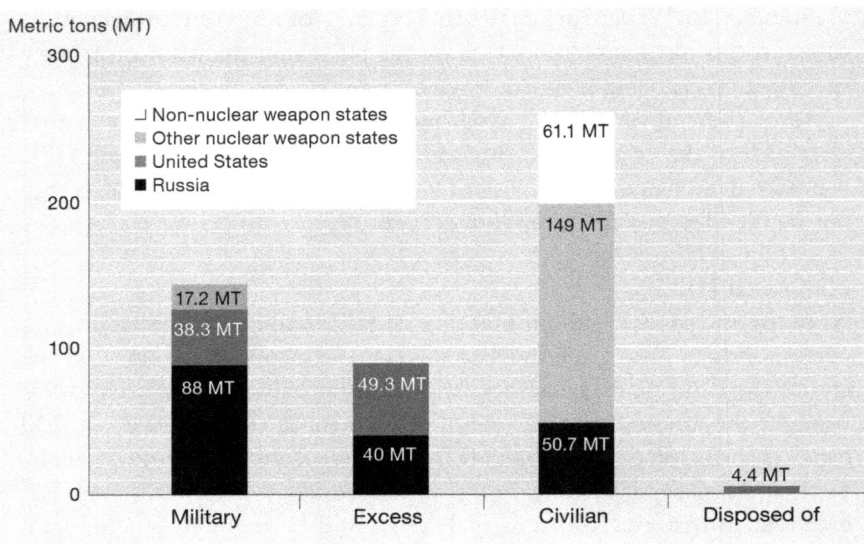

Metric tons (MT)

- Non-nuclear weapon states
- Other nuclear weapon states
- United States
- Russia

Military: 88 MT, 38.3 MT, 17.2 MT
Excess: 40 MT, 49.3 MT
Civilian: 50.7 MT, 149 MT, 61.1 MT
Disposed of: 4.4 MT

Figure 4.3
Global stockpiles of separated plutonium by category as of the end of 2012. Military stocks of plutonium are IPFM estimates with the exception of the inventories of the United States and the United Kingdom, which are based on their public declarations. Civilian stocks of separated plutonium are taken from INFCIRC/549 declarations where available. The uncertainty in the global military stockpile of separated plutonium is on the order of ±10 tons. The different stockpile categories shown in the figure are discussed in the text.

sufficient to account for considerable working stocks, only 50 tons of plutonium would be required for weapons. Accordingly, about 85 additional tons of plutonium beyond the material already declared excess to weapons purposes could be declared so. Since—unlike HEU, which is used in naval reactor fuel—there are no other military applications for plutonium, this material could be declared excess for any *military* purpose as well and brought under IAEA monitoring.

Additional Plutonium
About 10 tons of plutonium in Russia and India are difficult to categorize. After September 1994, when production for weapons officially ended, Russia continued to produce weapon-grade plutonium in three plutonium production reactors that supplied district heat and electricity to two local Siberian cities. Out of the estimated 15 tons produced between 1994 and 2010, only 9 tons are covered by the 2000

Russian-U.S. agreement on the disposition of excess weapons plutonium. The remaining 6 tons are being stored in oxide form and are subject to bilateral transparency measures to provide assurance that they will not be used in weapons. Russia has not included this material in its civilian stockpile declarations, however.

India also has plutonium stocks that are difficult to categorize. Under the 2008 U.S.-India nuclear deal, India has kept outside IAEA safeguards a growing stockpile of separated plutonium from spent heavy water power reactor fuel (about 5 tons as of 2012) and has categorized this material as "strategic." India plans to use this plutonium as fuel to start up its planned fleet of breeder reactors, which it has also kept outside international safeguards. In principle, natural uranium "blankets" around the cores of these breeder reactors could be used to turn this fuel-grade plutonium (about 70 percent plutonium-239) into an equivalent amount of weapon-grade plutonium (over 90 percent plutonium-239).[17]

These Russian and Indian stocks of material are neither clearly military nor civilian and unnecessarily add to the uncertainties associated with fissile material stocks and use. To resolve this ambiguity, Russia and India could declare these stocks excess for weapons purposes and offer them for international safeguards.

Excess Weapons Plutonium

Following the post-Cold War reductions in their nuclear arsenals, three countries have declared a combined total of more than 90 tons of plutonium excess: the United States (53.7 tons),[18] Russia (34 tons), and the United Kingdom (4.4 tons). France has downsized its stockpile of nuclear weapons by about 50 percent since the end of the Cold War but it has thus far not declared any plutonium excess for military purposes.

There are different possible disposal options for plutonium. It can be used as reactor-fuel or directly disposed of as waste. In either case, disposal is difficult and costly and little progress has been made since the 1990s. Details of available options and ongoing efforts are discussed in chapter 9.

Civilian Plutonium

The separation of plutonium from spent fuel, and the associated stockpiling of this material, creates the gravest security risk from civilian nuclear power. The global stockpile of civilian plutonium as of the end of 2012 was about 260 tons—enough for 30,000 Nagasaki-type nuclear explosives. The United Kingdom, France, Russia, and Japan have accumulated the largest stocks of civilian plutonium, together accounting for over 90 percent of this inventory. Most of Japan's stockpile is held in France and the United Kingdom, where it was separated from Japanese spent fuel. Germany has been successfully drawing down its stockpile of separated plutonium since it stopped sending spent fuel abroad for reprocessing in 2005. The civilian plutonium stockpile will increase, however, if China, India, Japan, and Russia go forward with planned large-scale reprocessing programs. The future of Japan's reprocessing program is uncertain in the wake of the March 2011 nuclear disaster at Fukushima.

Reducing Stockpile Uncertainties: Toward Nuclear Transparency

The uncertainties in the global fissile material stockpile are large. Combined, they are equivalent to several thousand nuclear warheads. As nuclear arsenals are reduced, these uncertainties could become obstacles to further reductions. The error bars on Russia's HEU and plutonium stockpiles are the largest, but the 20–30 percent uncertainties in the estimates for France and China also are significant. It will be difficult to improve independent estimates such as those discussed in chapter 3 without additional information being made public.

Some governments may have better estimates of other weapon states' stockpiles of fissile material based on intelligence gathered over the years. For example, ongoing and cumulative plutonium production can be estimated by local and global krypton-85 concentrations in the atmosphere. The United States and perhaps other countries collected and analyzed this data on a continuous basis during the Cold War.

To reduce uncertainties in national fissile material inventories to a level that would be necessary to support and facilitate deeper cuts in the nuclear arsenals, increased openness and bilateral or multilateral cooperative verification will be required. The task will be challenging. Every nuclear weapon program has initially been shrouded in secrecy. But there has been an increase in transparency over time, especially since the end of the Cold War. Nuclear transparency has focused mostly on the sizes of the nuclear arsenals, and all five NPT weapon states (the United States, Russia, the United Kingdom, France, and China) have made public statements on some aspect of their nuclear weapon holdings.

Since the weapon states that are parties to the NPT have ended their production of fissile materials for weapons purposes, declarations of fissile material stocks by those countries among them that have not already done so would be particularly meaningful as a next step toward more transparency.[19] The declarations could serve as "baseline declarations" for additional transparency steps later on. The weapon states should begin compiling data for such declarations even if this information will not initially be made public. Experience has shown that preparing declarations becomes more difficult over time as production records deteriorate or become more difficult to interpret and as the experts who were actively involved in these programs retire.[20]

Some important precedents for fissile material declarations exist. As discussed in chapter 3, the United States has made detailed declarations of the history of both its HEU and plutonium stockpiles, including production by year and site, and has updated and refined these declarations over the years. The United Kingdom has declared its total stocks of military fissile materials but has provided little information about their production histories. Russia, France, and China have not made public any information on their military fissile material stocks. As noted earlier, however, the NPT weapon states declare their civilian plutonium stocks annually and publicly through the IAEA, with the United Kingdom and France also declaring stocks of civilian HEU.

Building on these precedents, all nuclear weapon states could publish, and commit to update annually, their complete holdings of HEU and plutonium. Weapon states could also offer breakdowns of their stocks by purpose: nuclear weapons, naval reactor fuel, and civilian use. This would strengthen confidence in the weapon states' commitment to openness and a verifiable disarmament process.

Deeper reductions in existing nuclear warhead and fissile material stockpiles will require still finer-grained declarations to foster confidence and facilitate verification. These could include detailed data on the history of HEU and plutonium production by year and facility. Ideally they also would include basic design information for existing and former production facilities.

The economic cost of implementing nuclear transparency measures of the kind suggested here is not likely to be high in comparison to the price already being paid by the weapon states to manage their nuclear weapons and fissile material complexes and legacy facilities. France, for instance, is committed to spending 500 million euros to dismantle its three plutonium production reactors at Marcoule and about five billion euros to dismantle the associated UP-1 reprocessing plant that separated plutonium for weapons.[21]

Increasing transparency will be politically challenging, however. China, India, Israel, North Korea, and Pakistan traditionally have maintained a policy of opacity about their nuclear weapons and fissile material stocks, while Russia has reverted to a high level of secrecy about nuclear issues after a decade of relative openness following the collapse of the Soviet Union. Increased transparency by these states will come about more slowly than for other weapon states. As a way to build confidence, transparency measures could be implemented step-by-step. By making a declaration, a state would neither be committing itself to subsequent increases in transparency about its nuclear weapons program nor accepting constraints on its nuclear arsenal.

Verifying Past Production of Fissile Materials

As important and welcome any unilateral and voluntary transparency steps are and would be, in the longer term, cooperative approaches will be required to increase confidence in the correctness and completeness of fissile material declarations by weapon states. Ultimately, this is likely to involve onsite inspections at declared production facilities by international, multilateral, or bilateral teams.

One particular approach that will become available once weapon states are ready to provide access to former production sites is "nuclear archaeology," which uses nuclear forensic techniques to analyze trace impurities in structural materials or in waste materials at former production sites to provide independent checks on declarations of historical plutonium production and uranium enrichment.

As an important example of what can be done, between 1992 and 1998, the U.S. Pacific Northwest National Laboratory conducted a research and development program to evaluate and develop the technical basis for estimating the cumulative production of plutonium in a graphite-moderated plutonium production reactor from a forensic analysis of trace elements in the graphite (figure 4.4).[22] The United States, Russia, the United Kingdom, France, China, and North Korea have used such reactors for producing weapon plutonium. Estimates derived in this way would be uncertain by a few percent but also largely independent of records provided by the host country. Nuclear archaeology approaches are needed for other kinds of facilities used for fissile material production and to recover useful forensic information from associated wastes.

It works to the advantage of nuclear archaeology that large quantities of materials and extensive facility operations are needed for every kilogram of fissile material produced.

Figure 4.4

Workers taking graphite samples from the Hanford C reactor in 1994 in an early effort to test a nuclear archaeology technique. A full-scale exercise was later carried out to estimate total plutonium production in a British Magnox reactor in Wales, and irradiated graphite samples were analyzed from U.S., Russian, and French reactors.

Source: Jim Fuller and U.S. Department of Energy.

The window of opportunity for nuclear archaeology is closing for some facilities, however. For example, most of the global stockpile of highly enriched uranium was produced in gaseous diffusion plants. These plants are all shut down and are being decommissioned and dismantled. In some cases, critical components like the diffusion barriers that could carry evidence of the operating history of the plant are being buried or scrapped and their metal recycled. The depleted uranium associated with U.S. production of weapon-grade uranium is also now being processed to extract more uranium-235 or prepared for disposal. This will make verification of how much HEU was produced by these plants more difficult, if not impossible.

To facilitate future verification of declarations of fissile material production, weapon states should commit to catalog and preserve operating records and sample materials. Priority should be given to facilities facing decommissioning and waste materials scheduled for disposal or processing. Weapon states also could offer former production facilities as verification test beds and invite partners with similar facilities to join "site-to-site exercises" in verification approaches and techniques.

Collaborative verification projects could be effective confidence-building measures and could help sustain the disarmament process. In the first stage of reductions, uncertainties in the nuclear weapon stockpiles would not be of much concern but, as the United States and Russia move to lower levels and other nuclear weapon states join in multilateral disarmament steps, it will be critical that they gain confidence in the stockpile declarations. If doubts and mistrust persist among the parties about the real sizes of the stockpiles, further reductions will be more difficult.

Verifying the fissile material declarations—especially those of the United States and Russia—will be enormous undertakings. It will take many years for states to gain mutual confidence in the correctness and completeness of these declarations. Verification should therefore begin as far in advance of deeper reductions in the nuclear arsenals as possible.

───►

Figure 4.5 (Appendix)

National Stocks of HEU as of the end of 2012 (top). Russia's stockpile has been updated to the end of 2013, when the blend-down of excess HEU was completed. The numbers for the United Kingdom and United States are based on official publications and statements. The civilian HEU stocks of France and the United Kingdom are based on their public declarations to the IAEA. Numbers with asterisks are IPFM estimates, often with large uncertainties. A 20 percent uncertainty is assumed in the figures for total stocks in China and for the military stockpile in France, about 30 percent for Pakistan, and about 40 percent for India. HEU in non-nuclear weapon (NNW) states is under IAEA safeguards.

National stocks of plutonium as of the end of 2012 (bottom). Civilian stocks are based on the INFCIRC/549 declarations and are listed by ownership, not by current location. Weapon stocks are based on IPFM estimates except for the United States and United Kingdom, whose governments have made declarations. Uncertainties in estimated military stockpiles for China, France, India, Israel, North Korea, Pakistan, and Russia are on the order of 10–30 percent. The plutonium India separated from spent heavy water power reactor fuel has been categorized by India as "strategic" and not to be placed under IAEA safeguards. Russia has 6 tons of weapon-grade plutonium that it has agreed to not use for weapons but not declared excess. As of 2009, the United States has disposed of 4.4 tons of excess plutonium as waste in its underground Waste Isolation Pilot Plant in New Mexico.

Appendix: National Stockpiles of Fissile Materials

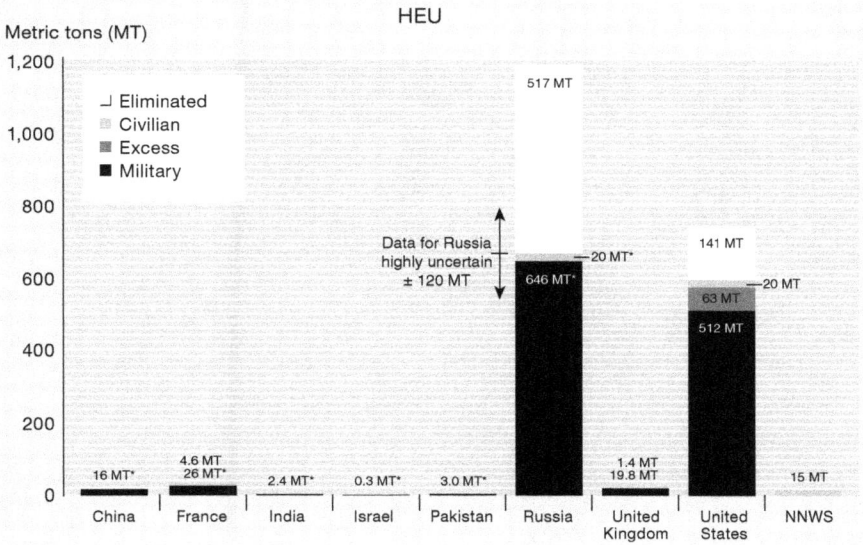

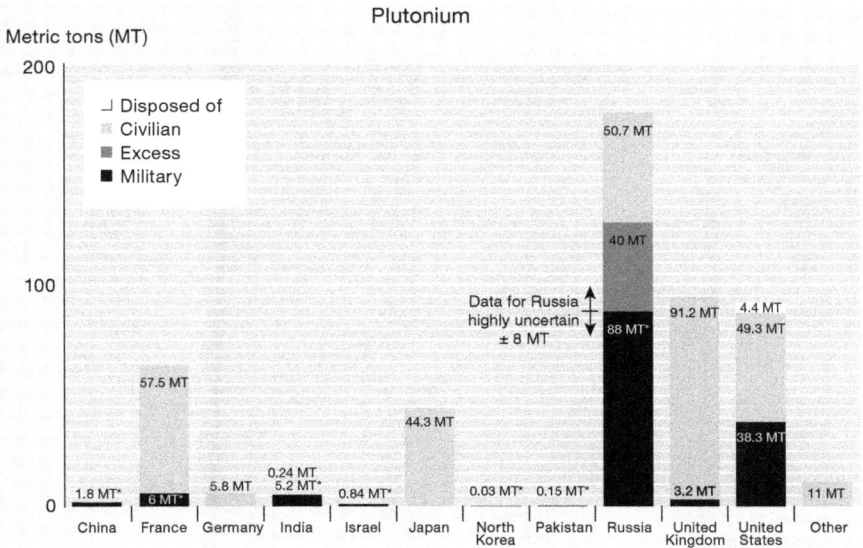

HEU

Metric tons (MT)

- ⌐ Eliminated
- ▨ Civilian
- ▪ Excess
- ■ Military

517 MT

Data for Russia highly uncertain ± 120 MT

—20 MT*

646 MT*

141 MT

—20 MT

63 MT

512 MT

1.4 MT
19.8 MT

16 MT*   4.6 MT
26 MT*   2.4 MT*   0.3 MT*   3.0 MT*   15 MT

China   France   India   Israel   Pakistan   Russia   United Kingdom   United States   NNWS

Plutonium

Metric tons (MT)

- ⌐ Disposed of
- ▨ Civilian
- ▪ Excess
- ■ Military

50.7 MT

40 MT

Data for Russia highly uncertain ± 8 MT

88 MT*

91.2 MT   4.4 MT
49.3 MT

57.5 MT

44.3 MT

38.3 MT

1.8 MT*   6 MT*   5.8 MT   0.24 MT
5.2 MT*   0.84 MT*   0.03 MT*   0.15 MT*   3.2 MT   11 MT

China   France   Germany   India   Israel   Japan   North Korea   Pakistan   Russia   United Kingdom   United States   Other

II
Breaking the Nuclear Energy-Weapons Link

5

Fissile Materials, Nuclear Power, and Nuclear Proliferation

The relationship between nuclear power and nuclear proliferation is complex and has changed over time, but it was the U.S. 1953 Atoms for Peace program that launched the large-scale sustained dissemination of nuclear expertise and technologies to non-weapon states. Atoms for Peace also led to the establishment in 1957 of the International Atomic Energy Agency, with its dual mission of promoting the peaceful use of nuclear technologies and materials while, at the same time, monitoring their use with "safeguards" to ensure that they are not used to produce nuclear weapons.[1] This approach subsequently became the foundation for the verification requirements on non-weapon state parties to the 1970 Non-Proliferation Treaty and the modern nonproliferation regime.

The spread of nuclear power programs, however, also has led to the spread of sensitive enrichment and reprocessing technologies. This has given some non-weapon states the means of producing fissile material and thereby a "latent" proliferation capability, where a country could quickly produce nuclear weapons should it decide to do so. In the view of the former Director General of the IAEA, Mohamed ElBaradei,[2] "in 1970, it was assumed that relatively few countries knew how to acquire nuclear weapons. Now, with thirty-five to forty countries in the know by some estimates, the margin of security under the current non-proliferation regime is becoming too slim for comfort."

As of 2013, the future of civilian nuclear power is uncertain. Opposed to concerns about safety, cost, radioactive waste disposal, and proliferation, there are arguments that expanded reliance on nuclear power could make a significant contribution to mitigating global climate change. Any large-scale expansion of nuclear power would likely mean adoption of this technology by many more than the 31 countries that had it in 2013. Whatever the relative importance of nuclear power in the future, it will be critical to maximize the barriers between civilian nuclear power and nuclear weapons. Even a small nuclear power program can provide a nuclear weapon breakout potential.

In principle, designing nuclear power to be "proliferation resistant" could make it more difficult to exploit for acquiring nuclear weapons, provide international early warning if a country moves to do so, and help prevent terrorist groups from obtaining nuclear weapons. To do so may require giving up some national control of nuclear technologies and moving to a regime in which nuclear power is governed by a globally agreed set of rules that apply equally to all states.

Atoms for Peace and the Spread of Nuclear Technology

In 1945, with the end of World War II, the United States sought to prevent the spread of the nuclear weapon capabilities that it had developed during the war. This effort even included ending nuclear collaboration with its closest wartime ally, the United Kingdom, and trying to establish control over the world's uranium resources—despite warnings from scientists that these efforts would prove futile.[3]

The United States decided to pursue a fundamentally different strategy once its nuclear weapon monopoly was broken by the Soviet Union's first nuclear test in 1949, followed in 1952 by the first British nuclear test. In December 1953, President Eisenhower outlined this new approach in his "Atoms for Peace" speech before the General Assembly of the United Nations. He proposed the following:

> The Governments principally involved . . . begin now and continue to make joint contributions from their stockpiles of normal uranium and fissionable materials to an international Atomic Energy Agency. We would expect that such an agency would be set up under the aegis of the United Nations. . . . The Atomic Energy Agency could be made responsible for the impounding, storage, and protection of the contributed fissionable and other materials. . . . The more important responsibility of this Atomic Energy Agency would be to devise methods whereby this fissionable material would be allocated to serve the peaceful pursuits of mankind. . . . A special purpose would be to provide abundant electrical energy in the power-starved areas of the world.[4]

According to C. D. Jackson, Eisenhower's Special Assistant for Cold War strategy and the key drafter of the Atoms for Peace Speech, one publicly unstated but privately acknowledged goal of the new initiative was "to associate the United States with the cause of peace, the improvement of living standards and better health."[5] Sharing American nuclear technology was to be a diplomatic tool in the Cold War against the Soviet

Union (figure 5.1). The initiative led to two important international con-
ferences on the peaceful uses of atomic energy, held in Geneva in 1955
and 1958, where the United States and other countries with active
nuclear programs released a large trove of information about nuclear
technology.

By the end of 1958, the United States had concluded agreements for
nuclear cooperation, including bilateral safeguards, with 22 countries.
This cooperation included technical training programs at U.S. laboratories for foreign scientists on a wide range of topics. According to a 1979
report by the U.S. government's General Accounting Office, between
1959 and 1965 the Oak Ridge School on Nuclear Reactor Technology
hosted 115 foreigners from 26 countries and the Argonne International
School of Nuclear Science and Engineering hosted 413 from 44 countries.[6] All told, by the early 1970s, the United States had trained several
thousand foreigners in a wide range of nuclear engineering topics.[7] This
pattern in nuclear cooperation was soon followed by Canada, the United
Kingdom, France, and the Soviet Union. By the late 1970s, at least 24
countries were providing nuclear training.[8]

Assistance provided as part of the Atoms for Peace programs included
the provision of research reactors, often provided at no cost, as a step-
ping-stone toward a first power reactor. The implicit assumption was
that recipient states would later purchase power reactors from the states
that had supplied their training and research reactors. The supplier states
that were weapon states also expected to sell fuel for the power reactors
they provided, enriched in plants originally built for their weapons programs. Canada, which did not have enrichment capabilities, provided
heavy water reactors, which can be fueled with natural uranium.

Training abroad and work with imported research reactors led to the
development of indigenous capabilities to make nuclear weapons material in some countries. Munir Ahmad Khan, who trained at Argonne in
the 1950s and later became the head of Pakistan's nuclear weapons
program, explained:

The Pakistani higher education system is so poor, I have no place from which to
draw talented scientists and engineers to work in our nuclear establishment. We
don't have training system for the kind of cadre we need. But, if we can get
France or somebody else to come and create a broad nuclear infrastructure, and
build these plants and these laboratories, I will train hundreds of my people in
ways that otherwise they would never be able to be trained. And with that training, and with the blueprints and the other things that we'd get along the way,
then we could set up separate plants that would not be under safeguards, that

Figure 5.1

Atoms for Peace. Traveling exhibitions promoting the peaceful uses of nuclear energy were sent across the United States and to many countries as part of the Atoms for Peace program. The high level of international interest was evident in the over 1,400 delegates from 73 countries that participated in the 1955 Geneva Conference, along with almost as many observers and close to 1,000 journalists.[a]

Source: U.S. Department of Energy.

[a]John Krige, "Atoms for Peace, Scientific Internationalism, and Scientific Intelligence," in John Krige and Kai-Henrik Barth, eds., *Global Power Knowledge: Science and Technology in International Affairs*, Osiris, vol. 21 (Chicago: University of Chicago Press, 2006): 161–181.

would not be built with direct foreign assistance, but I would now have the people who could do that. If I don't get the cooperation, I can't train the people to run a weapons program.[9]

The Atoms for Peace program and the spread of nuclear power also led directly to the spread of reprocessing technology. During the first Geneva conference in 1955, French experts published detailed technical papers on plutonium separation techniques in order to challenge what they saw as "industrial secrecy" imposed by the weapon states (France was not yet a weapon state), "obliging the other nuclear powers to follow suit and likewise lift their secrecy."[10] During the second Atoms for Peace conference in 1958, France tried to get agreement for a similar release of sensitive details on uranium enrichment but did not succeed, largely because of U.S. opposition.

The principal release of information on uranium enrichment technologies—notably the gas centrifuge—came later. Gernot Zippe, an Austrian physicist, was captured by the Soviet Union in 1945 and sent to a camp where he worked with other scientists on the gas centrifuge. The machine he and collaborators developed became the inspiration for most future centrifuge programs.[11] In 1956, Zippe left the Soviet Union for West Germany, where he provided design information about the Soviet gas centrifuges. He was invited to the United States first in 1957 and then to the University of Virginia in 1958 where he reconstructed the Soviet centrifuge for the U.S. Atomic Energy Commission (AEC).

At the University of Virginia, Zippe wrote a technical report describing the design details of the machine, which was published by Oak Ridge National Laboratory as an unclassified report, ORO-315, in 1960.[12] The Soviet Union, the United States, the Netherlands, the United Kingdom, and Germany had already begun development of gas centrifuges, but Zippe's report was critical in launching development efforts in additional countries. The historical record shows that in the fifteen years after the AEC published ORO-315, at least nine additional countries produced centrifuges similar to Zippe's University of Virginia machine.[13] Centrifuge technology later spread also to Pakistan, Iran, Libya, and North Korea through black market operations linked to A. Q. Khan.[14]

There was a parallel spread of reprocessing capabilities, even though reprocessing technology is not part of the "once-through" fuel cycle used in most nuclear countries. Interest in reprocessing derived primarily from two sources: interest in acquiring a nuclear weapon option, and the dream of self-sustaining plutonium breeder reactors (chapter 6). Starting in the early 1970s, a number of non-weapon states acquired or pursued reprocessing technology, including Argentina, Belgium, Brazil, Germany, India, Italy, Japan, Norway, Pakistan, North Korea, South Korea, and Taiwan.[15] Ultimately, the technical capabilities transferred by Atoms for Peace programs would help support the exploration of nuclear weapons options in over twenty countries.[16]

By the mid-1960s, controlling the spread of nuclear technologies on an *ad hoc* basis appeared more and more questionable. Already in 1963, U.S. President John F. Kennedy, in anticipation of the first Chinese nuclear test, stated, "I am haunted by the feeling that by 1970 . . . there may be 10 nuclear powers instead of 4, and by 1975, 15 or 20."[17] The Soviet Union was similarly alarmed, and the two countries began to push for the negotiation and then entry into force of the Treaty on the Non-Proliferation of Nuclear Weapons. After India's first nuclear test in 1974, the leading nuclear states also organized a Nuclear Suppliers Group and agreed on restrictive conditions for transfers of enrichment and reprocessing technologies.

The IAEA and the Non-Proliferation Treaty

While it made available nuclear technologies to a large number of countries, President Eisenhower's Atoms for Peace initiative also laid the basis for the development of the current nonproliferation regime. In particular, it led directly to the establishment in 1957 of the International Atomic Energy Agency, an autonomous organization located in Vienna under United Nations auspices with the dual charges of assisting in the development of civilian nuclear energy programs and safeguarding these programs to make sure that they were not used to provide fissile materials for weapons.

The NPT, which was opened for signature in 1968 and entered into force in 1970, provides the legal framework for addressing nuclear proliferation. The treaty defines two classes of states: nuclear weapon states that "manufactured and exploded" a nuclear weapon before 1967 (the United States, the Soviet Union/Russia, the United Kingdom, France, and China), and non-weapon states.[18] The United States, Soviet Union, and United Kingdom signed the NPT in 1968, but it was only in 1992 that China and then France joined the treaty.

As of 2013, 190 states had joined the NPT.[19] Israel, India, and Pakistan are the only three states that never joined the NPT. North Korea withdrew from the treaty in 2003. Barring an amendment to the NPT, its definition of "nuclear weapon state" rules out the possibility of any additional state joining the treaty as a nuclear weapon state. The only option to join the treaty for such a state is to renounce its nuclear weapons and sign the NPT as a non-weapon state. This option was taken by South Africa in 1991 and also by Belarus, Kazakhstan, and Ukraine, countries that inherited nuclear weapons from the Soviet Union.[20]

In Article I of the NPT, the nuclear weapon state parties to the treaty undertake not to transfer nuclear weapons to any other country. In Article II, the non-weapon states undertake not to acquire nuclear weapons and, in Article III, they agree to accept IAEA safeguards on their peaceful nuclear programs to assure that no diversions are taking place. At the same time, however, Article IV of the treaty states the following:

1. Nothing in this Treaty shall be interpreted as affecting the inalienable right of all the Parties to the Treaty to develop research, production, and use of nuclear energy for peaceful purposes without discrimination and in conformity with Articles I and II of this Treaty.
2. All the Parties to the Treaty undertake to facilitate, and have the right to participate in the fullest possible exchange of equipment, materials, and scientific and technological information for the peaceful uses of nuclear energy.[21]

The NPT requires the acceptance of IAEA safeguards only by non-weapon states. Two of the NPT weapon states, France and the United Kingdom, are members of the 1957 Euratom Treaty, which covers the European Union and requires—even of weapon-state member countries—that all materials declared to be civilian be placed under Euratom safeguards. The IAEA oversees and complements Euratom controls of nuclear materials in the European Union.

The five NPT weapon states have offered to make some of their civilian nuclear facilities eligible for IAEA safeguards. In practice, however, virtually none of these facilities have been selected by the IAEA due to budgetary constraints and its policy to prioritize inspections in non-weapon states. In the cases of India, Pakistan, and Israel, some imported facilities are under IAEA safeguards as a result of requirements imposed by the countries that supplied those facilities.

The purpose of IAEA safeguards, which include a system of audits, surveillance, and inspections at declared nuclear facilities, is to verify that nuclear materials are not diverted to weapons purposes. The model comprehensive safeguards agreement that non-weapon states accept is described in IAEA Information Circular 153 (INFCIRC/153). This agreement obliges a state "to accept safeguards . . . on all source or special fissionable material in all peaceful nuclear activities within its territory, under its jurisdiction or carried out under its control anywhere, for the exclusive purpose of verifying that such material is not diverted to nuclear weapons or other nuclear explosive devices."[22] These are referred to as "full-scope" or "comprehensive" safeguards.

As of the end of 2012, the IAEA had safeguards agreements in force with 179 countries, and Taiwan, and carried out regular inspections at over 1,300 facilities worldwide.[23] This included seventeen facilities under the facility-specific INFCIRC/66 safeguards in India, Israel, and Pakistan, which are not members to the NPT, and twelve facilities under the voluntary offers made by the NPT weapon states. The annual IAEA safeguards budget was about $200 million.[24]

The specific goal of safeguards is to detect and provide timely warning of the diversion of a significant quantity (SQ) of fissile material from non-weapons uses. The IAEA defines an SQ as the "approximate amount of nuclear material for which the possibility of manufacturing a nuclear explosive device cannot be excluded." The definition allows for losses associated with warhead component manufacturing.

The NPT weapon states have advised the IAEA that appropriate values for SQs are:

• 8 kilograms of plutonium of virtually any isotopic composition,

• 8 kilograms of uranium-233, and

• 25 kilograms of uranium-235 contained in HEU.

The timeliness goals that the safeguards are designed to achieve for detecting diversion are an estimated seven to ten days for a country to convert metallic plutonium, HEU, or uranium-233 metal to weapon components; one to three weeks to convert unirradiated compounds (such as oxides) of plutonium, HEU, or uranium-233; and one to three months to extract and convert to metal weapon components plutonium, HEU, or uranium-233 in irradiated fuel.[25]

The Gulf War of 1991 revealed that Iraq, a party to the NPT, had been pursuing a clandestine nuclear weapons program and been able to skirt the IAEA safeguards system. This led in 1997 to a proposed confidence building arrangement, in the form of an Additional Protocol (INFCIRC/540) to national safeguards agreements with the IAEA that would give the IAEA tools to look for undeclared nuclear activities.[26]

The Additional Protocol is voluntary: parties to the NPT do not have to accept it, though once accepted it is legally binding. The protocol has become, however, a de facto norm for all NPT parties including the NPT weapon states. Under the Additional Protocol, non-weapon-state parties must provide the IAEA with greater information about and access to a country's uranium mines, nuclear waste sites, and nuclear research and development facilities. If approved by the IAEA's Board of Governors, the Additional Protocol would also allow the Agency to use wide-area environmental sampling along with its regular safeguards measurements to look for undeclared nuclear activities. As of the end of 2013, 122 states had Additional Protocols in force with the IAEA.[27]

Concern about the spread of nuclear technology, materials, and expertise has also included the fear that nations or subnational terrorist or criminal groups might steal fissile materials to make nuclear weapons. The discovery in 1965 of the loss and possible diversion of large amounts of weapon-grade uranium from the U.S. Nuclear Materials and Equipment Corporation (NUMEC) facility in Apollo, Pennsylvania, led the U.S. Atomic Energy Commission to set up in July 1966 an Advisory Panel on Safeguarding Special Nuclear Material.[28] The panel's report in 1967 called for "minimum physical protection standards" for fissile material to prevent theft, arguing that, along with seeking to prevent non-weapon states from diverting material from civilian nuclear programs for weapons purposes, "safeguards programs should also be designed in recognition of the problem of terrorist or criminal groups clandestinely acquiring nuclear weapons or materials useful therein."[29]

Lack of progress in developing and implementing effective physical security led to more public expressions of concern in the 1970s, including by former leading nuclear weapon designer Theodore Taylor.[30] Taylor detailed the risks of nuclear materials theft in the United States and the relative ease with which a small group, and possibly even an individual, could design and fabricate a simple nuclear explosive.[31]

In the early 1990s, a new phenomenon emerged and, for a time, became a major concern in the security community: gram and, in some cases, kilogram quantities of nuclear material were intercepted, primarily in Europe, pointing to losses and thefts facilitated by weak security arrangements and the possibility of an existing black market for this material. Among the most significant cases was the interception of 0.36 kilograms of plutonium at Munich Airport in August 1994 on a Lufthansa flight from Moscow; and later that year, in December 1994, 2.72 kilograms of 88 percent enriched HEU was seized from the back of a car parked in Prague. The origin of these materials could often be traced back to the former Soviet Union, but in many cases the original source or the point at which national control was lost could not be determined.

Since 1995, the IAEA has maintained a database of incidents of *confirmed* cases of illicit trafficking[32] covering all types of nuclear materials including naturally occurring and artificially produced radioisotopes and radioactively contaminated material such as scrap metal. As of 2013, the database had more than 2,000 confirmed entries and included sixteen cases involving highly enriched uranium and plutonium.[33] The frequency of intercepts has not increased over the years, and there is a widely shared view that the security of nuclear materials has significantly improved since the collapse of the Soviet Union.

The September 11, 2001, attacks on the United States increased concerns about the possibility of nuclear terrorism and led in April 2004 to UN Security Council Resolution 1540 requiring all states to "adopt and enforce appropriate effective laws which prohibit any non-State actor to manufacture, acquire, possess, develop, transport, transfer or use nuclear, chemical or biological weapons and their means of delivery."[34] States are also required "to develop and maintain appropriate effective measures to account for and secure" nuclear, chemical, or biological weapons and materials, including putting in place physical protection measures, border controls, law enforcement efforts to prevent illicit trafficking, and export and transshipment controls. The resolution required each state to submit, within six months, a report on what measures it had taken to comply. A "1540 Committee" was established to oversee implementation of the resolution. In April 2011, the Security Council extended the mandate of the 1540 Committee until 2021. As of the end of 2012, 169 states had submitted reports on their implementation of the Resolution, while twenty-four states had not yet made their first submission.[35] Even among reporting states, however, there remained issues of a lack of effective implementation due in some cases to a low priority accorded to the issue and in other cases the need for technical and financial assistance.[36]

Proliferation Resistance

Throughout most of the nuclear era, projections of future nuclear growth have been consistently too optimistic (figure 5.2). Growth in electricity demand slowed due to the end of price declines after the 1973 Arab oil embargo. Many projects were abandoned or suspended indefinitely, and licensing and construction problems delayed completion of other nuclear power plants by many years. Also, as the costs of nuclear power plants

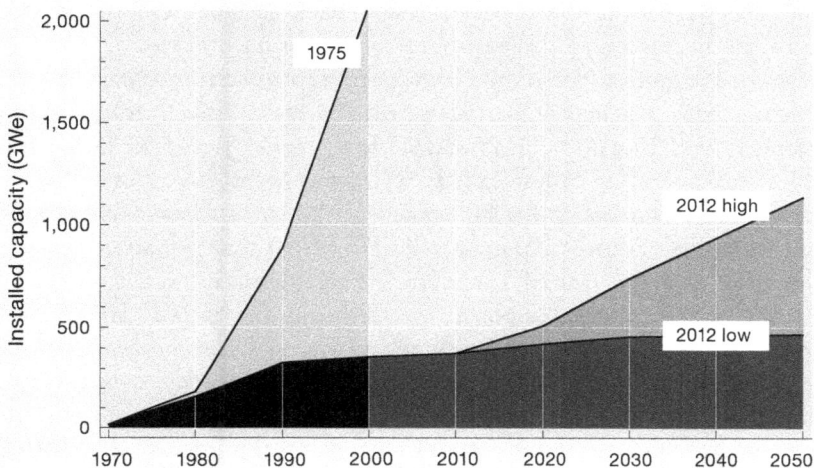

Figure 5.2

IAEA forecasts made in 1975 and 2012 for global nuclear capacity growth. Projections made in the 1970s for the growth of nuclear power were used to argue for an early large-scale deployment both of light water and breeder reactors and proved to be much too high. More recent projections suggest a more limited role for nuclear power in the future electricity market.

Data from R. B. Fitts and H. Fujii, "Fuel Cycle Demand, Supply and Cost Trends," *IAEA Bulletin* 18, no. 1 (1975): 19; and *Energy, Electricity and Nuclear Power Estimates for the Period up to 2050, 2012 Edition* (Vienna: International Atomic Energy Agency, 2012).

skyrocketed in the United States and Europe, they became less attractive to electrical utilities. Public opposition to nuclear power also increased due to the 1979 reactor accident at Three Mile Island in Pennsylvania in the United States, the 1986 reactor meltdown at Chernobyl in Ukraine (then part of the former Soviet Union), and the March 2011 disaster at Fukushima in Japan. There have also been enduring public concerns about unresolved waste disposal issues.

Since 1990, there have been relatively few new construction starts of nuclear power plants with the exception of plants in Asia, notably in China and South Korea. The market share of nuclear power has been dropping since 2001 because of the faster growth of non-nuclear generation. Still, before the Fukushima accidents of March 2011, there had been an active debate about the possibility of a "nuclear renaissance," driven in part by concerns over climate change. Nuclear power and its fuel cycle, on a life-cycle basis, emits only a few percent as much carbon dioxide to the atmosphere per kilowatt-hour as coal or natural gas.

As of the end of 2013, the IAEA listed 435 operating nuclear power reactors with a generating capacity of 373 gigawatts-electric in thirty countries and Taiwan.[37] This included forty-eight reactors in Japan, but none were operating pending safety reviews after the Fukushima disaster. Nine countries account for 80 percent of global nuclear capacity: the United States, France, Japan, Russia, South Korea, Canada, Ukraine, Germany, and the United Kingdom. After the Fukushima accident of 2011, Germany decided to phase out nuclear power by 2022, and France and Japan planned to reduce their reliance on nuclear energy.

From the beginning of the nuclear era, there has been a debate about the extent to which the existence of a nuclear power program, and its attendant scientific capabilities and technologies, influences a country's proliferation decisions.[38] There is no doubt, however, that a civilian nuclear power program provides a state with a foundation for the production of fissile material. It allows a country to train scientists and engineers, acquire research facilities and nuclear power reactors, and possibly also to learn the techniques of reprocessing and enrichment that could later be turned to producing weapons materials. Even small civilian nuclear energy programs can involve large stocks and flows of plutonium and a large capacity to produce HEU when measured in weapon equivalents. A civilian program can thus carry a country down a path of latent proliferation, in which it moves closer to nuclear weapons without actually having to make an explicit decision to acquire them.[39]

There are many ways to deploy and use nuclear energy, differing especially with regard to the possession of nuclear fuel cycle facilities. At present, there are two principal power reactor types in operation: one using natural uranium fuel and heavy water as moderator (notably the Canadian CANDU reactor), of which there were about 50 units as of the end of 2013, and one using low-enriched uranium as fuel and light (ordinary) water as moderator (the light water reactor or LWR, of which there were about 350 units). In the future, other reactor types that are currently under development might be deployed in larger numbers including fast-neutron sodium-cooled breeder reactors, high temperature gas-cooled reactors, thorium-based reactors, and small modular reactors of various kinds. These will have differing proliferation-resistance characteristics to some degree, but the key concern is likely to remain the fuel cycle: whether or not the nuclear power system of a country is based on recycling plutonium or uranium-233 separated from spent fuel, and whether or not a country has nationally owned and operated facilities for reprocessing, uranium enrichment, or both.

As of 2013, every country is free in principle to deploy its preferred fuel cycle and to keep all relevant facilities under national control. Under the NPT, non-weapon states are required to accept IAEA monitoring, but a state intending to break out and produce nuclear weapons can minimize the interval between the IAEA's detection of the country's fissile material diversion and its acquisition of nuclear weapons. To map out how this warning time depends on a country's possession of different nuclear facilities, four classes of countries are considered below:

- with reprocessing plants;
- with uranium enrichment plants;
- with only reactors; and
- without nuclear power reactors.

Countries with Reprocessing Plants

Countries that operate reprocessing plants—or have available a stock of plutonium from the past reprocessing at home or in other countries of their spent fuel—can obtain plutonium for weapons almost immediately. The plutonium typically discharged from commercial LWRs is not "weapon-grade" (more than 90 percent plutonium-239) but is weapon-usable.[40]

Moreover, any country with an LWR could produce weapon-grade plutonium by discharging a batch of fuel after it had reached only about 10 percent of its design burnup. At that point, the 20-ton load of fuel typically discharged annually by a one gigawatt electric LWR would contain about 40 kilograms of weapon-grade plutonium, which is enough for more than five Nagasaki-type explosives.

In a country with a reprocessing plant, this plutonium could be separated without much delay. It might then take days or weeks to process the plutonium into weapon components.

Countries with Uranium Enrichment Plants

The dominant nuclear power reactor type, the light water reactor, uses low-enriched fuel (LEU) with a uranium-235 content of 4–5 percent, far less than the concentration needed for a weapon. However, a uranium enrichment plant can be used to produce weapon-grade uranium as well as LEU, and most nuclear weapon states have used their enrichment plants for both these purposes.

Centrifuge plants can easily be shifted from producing LEU to weapon-grade uranium. Plants designed to produce LEU are organized into cascades to convert natural uranium into LEU, and a commercial plant typically has many such cascades operating in parallel. The quickest path to produce weapon-grade uranium at a centrifuge facility designed to produce LEU could be through interconnection of existing cascades without reconfiguration of the cascades themselves.[41] In this manner, it would be possible within days to begin the production of weapon-grade uranium.[42] For a small enrichment facility sized to fuel a single one giga-watt-scale LWR, the time required to produce HEU sufficient for a few weapons would be on the order of a few weeks. Commercial centrifuge plants in operation are typically very much larger, however, sized to fuel tens of power reactors. A facility sized to support ten large power reactors would be capable of providing enough HEU for several weapons per week.[43]

Countries that have national enrichment plants therefore are virtual nuclear weapon states. In 2013, there were thirteen states with uranium enrichment plants, including the non-weapon states of the Netherlands, Germany, Brazil, Japan, and Iran (figure 5.3). (See appendix 1 for a list of all major operating uranium enrichment plants.) Several additional countries have demonstrated enrichment capability or have operated research or pilot-scale plants.

Countries with Once-Through Nuclear Fuel Cycles and No Reprocessing or Enrichment Plants

In most countries with nuclear power programs, no HEU is used and no plutonium is separated from spent fuel. Spent fuel from a commercial light water reactor typically contains about 1 percent plutonium, however. The eight kilograms of plutonium that the IAEA assumes is sufficient to make a first-generation Nagasaki-type bomb could therefore be recovered from a single ton of spent fuel.

To separate this plutonium, a country would have to reprocess the spent fuel. If a country were able to build in advance without detection a small "quick and dirty" reprocessing plant—that is, a plant with minimal radiation protection and rudimentary radioactive waste management—it could produce plutonium for weapons within months.

The possibility of the quick and secret construction of a reprocessing plant was first raised in a 1977 study by a group of technical experts at the U.S. Oak Ridge National Laboratory who were so upset by nonproliferation arguments against reprocessing that they developed and

Figure 5.3

In April 2008, Iran made public a collection of photos showing then President Mahmoud Ahmadinejad touring the Natanz enrichment plant (a). Some also showed details of centrifuge components revealing the status of Iran's enrichment technology (b). A plant sized to enrich enough uranium to make fuel for a one gigawatt-scale power reactor gives a country an effective breakout capability and could be reconfigured to make highly enriched uranium for twenty to thirty nuclear weapons per year.

Source: www.president.ir. These pictures were subsequently removed from the website.

distributed an unclassified design of a quick and dirty reprocessing plant together with a detailed flow sheet and an equipment list (figure 5.4). Their study suggested that such a plant might be constructed in a year or less.[44] A relatively small reprocessing plant with a capacity of 50 tons heavy metal per year could separate enough plutonium for a single bomb in about a week.

Countries without Nuclear Power

A state that had never had a nuclear energy program would take the longest time to produce fissile material for weapons purposes. It might have some nuclear engineering expertise and perhaps small research reactors for scientific, industrial, medical, or other civilian purposes. Such a country, if it wished to produce fissile material, would have to construct an enrichment plant to make highly enriched uranium or construct a dedicated production reactor and reprocessing plant to obtain plutonium. If it had a research reactor of sufficiently high power, it would need to construct only a reprocessing plant to obtain plutonium.

If a state wanted to develop in secret a nuclear weapon capability, it would probably choose gas centrifuge enrichment technology. Centrifuges can be deployed on a small scale, and a small centrifuge plant is easily hidden (figure 5.5). It does not need an unusual amount of electric power and leaks only a tiny amount of UF_6, making detection at any distance extremely difficult.[45] Historically, it has taken upward of a decade for a state to master gas centrifuge technology without outside assistance. It could take less time, however, if a program sought only to make simple centrifuges for weapons purposes without aiming for the high separation efficiency and throughput required to compete in the global uranium enrichment market.[46]

The plutonium route could, in principle, be quicker. Lacking spent fuel along with a reprocessing plant, a state would need to build a production reactor. Such a reactor was discovered to be under construction in Syria; it was bombed and destroyed by Israel in September 2007. The time to build a production reactor and reprocessing plant and irradiate and reprocess the fuel would be a few years—perhaps less for a state with greater fuel cycle expertise and the capability to produce or acquire the specialized components and materials normally associated with enrichment and reprocessing programs.

This analysis may also help answer the question of whether and under what conditions civilian nuclear power could be compatible with nuclear disarmament.[47]

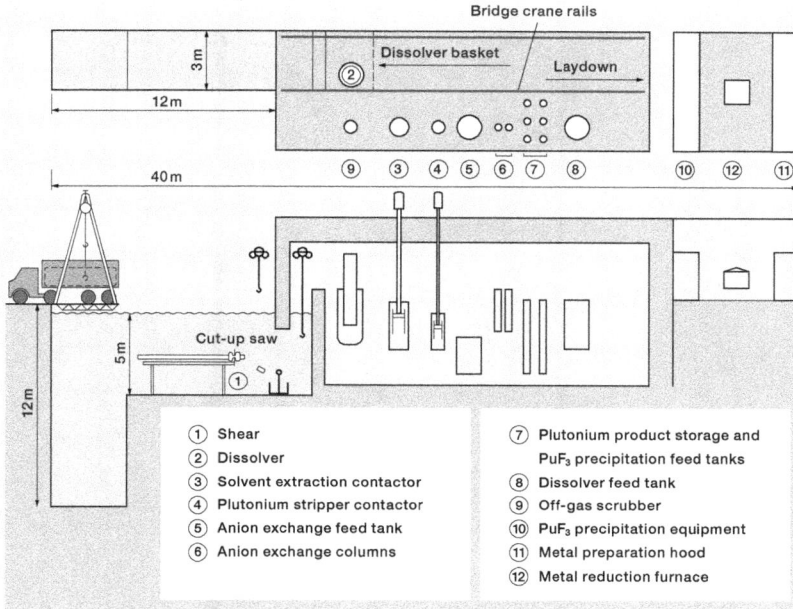

Figure 5.4

A "simple, quick reprocessing plant," based on the PUREX process (top view above, side view below). A reprocessing "canyon" is dug into the ground rather than built above ground between heavy radiation-shielding walls. The spent fuel would be delivered by truck in a massive radiation-shielding cask at the left. The cask would be lowered into a deep pool of water where its top would be removed and the irradiated uranium fuel lifted out with the water providing radiation shielding. The fuel would be cut up into short lengths under water (1) with the segments dropping into a basket that would carry them to a dissolver tank of hot nitric acid (2) where the uranium, plutonium, and fission products would be leached out of the fuel cladding. The resulting acid solution would then be pumped to a second tank (3) where it would be mixed with droplets of organic solvent that would pick up the plutonium and uranium and float to the top, leaving the fission products behind. The organic solution of plutonium and uranium then would be pumped to a third tank (4) where it would be mixed with dilute nitric acid that would pick up the plutonium, leaving the uranium behind. The plutonium solution would go through further cleanup in an ion exchange column (5, 6) and then be mixed with hydrofluoric acid to precipitate it out as solid plutonium trifluoride, PuF_3 (7, 8, 10). If adequately decontaminated PuF_3 were produced, it could be reacted with calcium in an aboveground glove box to remove the fluoride, leaving behind plutonium metal (11, 12).

Source: Adapted from D. E. Ferguson, "Simple, Quick Processing Plant, Intra-Laboratory Correspondence" (Oak Ridge National Laboratory, August 30, 1977).

Figure 5.5

Site of Iran's underground Fordow enrichment plant near the city of Qom. The partially built Fordow fuel enrichment plant (FFEP) was publicly disclosed on September 25, 2009, during a G20 summit in Pittsburgh, Pennsylvania. Iran declared its existence to the IAEA at about the same time. IAEA inspectors first visited the site in late October 2009 and confirmed that the plant could eventually hold about 3,000 centrifuges.
Source: Google Earth.

Reprocessing and uranium enrichment facilities under national control afford a country the quickest path to nuclear weapons, should it have the intention to acquire nuclear weapons or wish to have the option to do so. This is also true for a country seeking to break out of a disarmament agreement. At present, such facilities are located in only a few countries, most of which are already nuclear weapon states or—for at present, at least—have decided not to seek nuclear weapons. But, in the long term, a nonproliferation regime that forbids technologies and fuel cycles to some countries while allowing them in others will not be accepted. South Korea, in its negotiations of a new Agreement for Nuclear Cooperation with the United States, for example, has insisted on the same right to enrich and reprocess that the United States had agreed to for Japan in 1988.

For subnational groups, the most plausible route to fissile material would be to divert HEU fuel used in some research reactors or to divert separated plutonium from fuel cycles involving reprocessing and plutonium use as a reactor fuel in the form of a mixture of plutonium and uranium oxides. Fabricating nuclear weapons starting with fresh HEU or mixed oxide (MOX) fuel could be within the capabilities of a subnational group. For this reason, HEU and separated plutonium should be eliminated from civilian nuclear fuel cycles.

To meet concerns about national proliferation and the terrorist acquisition of nuclear weapons, as is discussed further in chapter 6, serious consideration should be given to the possibility of phasing out reprocessing plants altogether. In principle, this should not be a difficult decision, since reprocessing will not be necessary or economic for the foreseeable future.[48]

In the case of enrichment, replacing national enrichment plants with facilities under regional, multinational, and international control would reduce the danger of a country using a national enrichment facility to produce weapon-grade uranium.[49] Such a transition would take time but, at the least, it could be required that any new enrichment facilities be multinational. This raises questions about where such facilities would be built and by whom and what sort of management structure provides effective control over the use of such a facility.[50] Collective ownership works to create a political barrier to the host state seizing the plant for weapons purposes, but ultimately access to a fissile material capability will be available to states that have such plants sited on their territory.[51]

A phaseout of civilian nuclear energy would provide the most effective and enduring constraint on proliferation risks in a nuclear weapon-free world, but any policymaking process on a phaseout also will have to consider the costs and benefits of alternative energy policies.[52] These costs and benefits need to be better understood by policymakers and the public so as to allow the nuclear weapons debate to focus more clearly on the linkages between nuclear energy use, nonproliferation, the prevention of nuclear terrorism, and the path to total nuclear disarmament.

6

Ending the Separation of Plutonium

Plutonium was discovered in a series of experiments starting at the end of 1940 and its fission was demonstrated soon afterward in early 1941. It was quickly recognized that this artificial chain-reacting element could be a potential nuclear weapon material, and soon one major focus of the U.S. Manhattan Project became to build nuclear reactors to produce plutonium in weapon quantities. At the same time, some nuclear scientists hoped that some good might possibly come out of the nuclear weapons program—specifically that nuclear reactors could offer a large-scale source of electric power for society. After World War II, the United States, the Soviet Union, the United Kingdom, France, China, India, and countries inspired by their examples launched ambitious programs for the use of nuclear energy, including efforts to use plutonium as a reactor fuel.

Over the past sixty years, these civilian programs have created a legacy of about 260 tons of separated civilian plutonium. This stockpile poses two major global challenges: how to prevent it from being used for weapons and how to safely dispose of it.

Plutonium Breeder Reactors

The nuclear pioneers believed that, if power reactors only exploited efficiently the energy in the chain-reacting uranium isotope uranium-235 (0.7 percent of natural uranium), the amount of natural uranium available in rich deposits would not be sufficient to support a significant fraction of humanity's long-term energy needs. For this reason, in 1945, Enrico Fermi, Leo Szilard, and some of the other scientists gathered at the University of Chicago to design the first plutonium production reactors concluded that it would be necessary to develop reactors that would exploit the energy latent in non-chain-reacting uranium-238, which, nucleus by nucleus, contains as much potential fission energy as uranium-235 and is 140 times more abundant (99.3 percent of natural uranium). In theory, such reactors could release from the 3 grams of uranium in a ton of average crustal rock more energy than is released by burning a ton of coal.[1]

Szilard invented a plutonium "breeder" reactor that would be fueled by plutonium while producing from uranium-238 more plutonium than it consumed, thus making uranium-238 the ultimate fuel for the breeder.[2] An alternative breeder reactor based on the artificial chain-reacting element uranium-233 bred from thorium was later considered but abandoned because the theoretical rate of growth of its stockpile of fissile material would be less.[3]

Glenn Seaborg, co-discoverer of plutonium and other transuranic elements, for which he shared the 1951 Nobel Prize for Chemistry, chaired the U.S. Atomic Energy Commission from 1961 to 1971 and enthusiastically promoted a vision of a future global economy powered by plutonium.[4] Nuclear establishments around the world embraced this vision and set to work preparing for a transition to plutonium breeder reactors.

Although uranium enriched to about 20 percent uranium-235 could have been used to provide the initial fuel for breeder reactors,[5] nuclear planners generally assumed that the fissile fuel for starting up the first breeder reactors would be plutonium separated from the spent fuel of already existing uranium-235 fueled reactors. Thereafter, as the fraction of breeder reactors in the nuclear energy mix increased, an increasing fraction of new breeder reactors could be started up with the surplus plutonium from the breeders themselves. This surplus plutonium would be produced in a "blanket" of uranium around the breeder core that would capture in uranium-238 most of the neutrons leaking out of the core, converting the uranium-238 into uranium-239, which would then decay into plutonium-239 (figure 6.1). Plutonium would be produced in the breeder core as well, which would contain about 20 percent plutonium mixed with 80 percent uranium-238.

In a nuclear fuel cycle based on breeders, there would be continuous reprocessing of spent fuel to separate plutonium from first-generation reactor spent fuel and from irradiated breeder fuel and blanket assemblies. The plutonium would be separated from the highly radioactive fission products with which it was created behind thick radiation shielding in a remotely operated chemical spent fuel "reprocessing" plant. The separated plutonium then would be mixed with depleted uranium and fabricated into new breeder fuel.

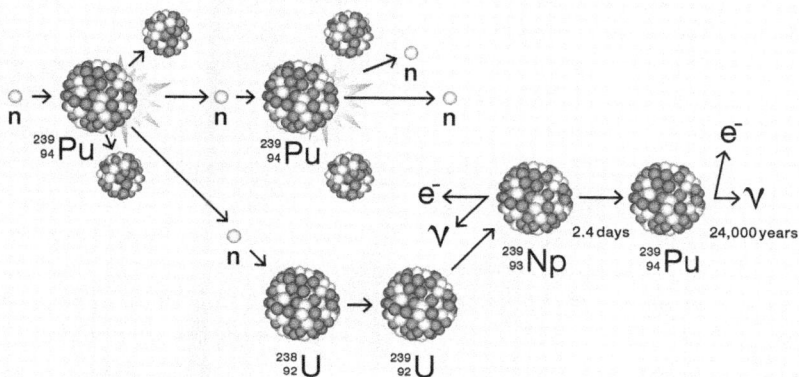

Figure 6.1

A breeder reactor releases energy from a fission chain reaction in plutonium while most of the extra neutrons released from the fissions convert uranium-238 into plutonium fuel. In contrast to slow-neutron water-cooled power reactors, fast-neutron sodium-cooled breeder reactors fueled with plutonium would be able to produce plutonium at a higher rate than they consumed it.

To maximize the number of new plutonium atoms produced per fission, it is necessary to maximize the number of neutrons produced per fission. Since plutonium fissioned with fast neutrons releases more neutrons on average than when it is fissioned with slow neutrons, it was decided to use a coolant that minimized the energy loss of the neutrons between fissions. This ruled out water, which slows neutrons.[6] It therefore was proposed that a metal with a heavy nucleus, a low melting point, a low affinity for capturing neutrons, and a low viscosity be used as a coolant.[7] Most R&D efforts focused on sodium, which has a melting temperature of 97 degrees centigrade and therefore would be liquid at well below the operating temperature of a power reactor.

The great challenge in designing and operating sodium-cooled reactors is that sodium burns upon exposure to air or water. Any leak of sodium into air or water therefore can result in severe damage. This greatly complicates reactor maintenance and refueling. The steam generators, where hot sodium flows through thin tubes surrounded by high-pressure water, also are challenging to design.

Hyman G. Rickover, who oversaw the development of propulsion reactors for the U.S. Navy, was initially attracted to fast-neutron sodium-cooled reactors because their cores are more compact than those of water-cooled reactors. Therefore, in 1955, he had a sodium-cooled reactor installed in the second U.S. nuclear-powered submarine, the *Seawolf*. Based on the problems encountered in this test he concluded, however, that sodium-cooled reactors are "expensive to build, complex to operate, susceptible to prolonged shutdown as a result of even minor malfunctions, and difficult and time-consuming to repair."[8] After a few years, the troubled *Seawolf* reactor was replaced with a water-cooled reactor.

During subsequent decades, roughly $100 billion (2010 $) were spent globally on attempts to commercialize sodium-cooled breeder reactors.[9] France, Germany, Japan, the Soviet Union/Russia, the United Kingdom, and the United States all built prototype and demonstration reactors. All but Russia's BN-600 operated for only a relatively few years, however, and, even during that period, for a small fraction of the time (table 6.1).[10] As of 2013, the BN-600 had operated for more than thirty years with a respectable capacity factor. It also had fourteen sodium fires in its first

Table 6.1
Performance of prototype and demonstration fast breeder reactors

Country	Reactor	Generating capacity (GWe)	Years of operation	Cumulative capacity factor through 2011 (%)[a]
France	*Phénix*	0.13	1973–2010	41
	Superphénix	1.2	1986–1996	8
Germany	*SNR-300*	0.33	None[b]	–
Japan	*Monju*	0.28	1995[c]	–
Russia	*BN-600*	0.56	since 1980	74
United Kingdom	*Prototype Fast Reactor*	0.25	1978–1991	27
United States	*Fermi I*	0.061	1966–1972[d]	0.6
	Clinch River	0.35	None[e]	–

Note: GWe = gigawatts electric.
[a]The capacity factor is the ratio of the amount of electrical energy that a power plant produces to that it would have produced had it operated at full capacity all the time. Globally, the capacity factor for water-cooled nuclear power plants has averaged about 80 percent since the late 1990s.
[b]Completed in 1985 and officially canceled in 1991. It was never operated because of safety concerns.
[c]Shut down by a sodium fire after three months.
[d]Shut down after a partial meltdown.
[e]Canceled in 1983 before construction started.

seventeen years.[11] No breeder reactor has yet been built that is economically competitive with water-cooled reactors.

In the meantime, the anticipated crisis in the availability of uranium receded beyond any reasonable planning horizon. In 1975, the IAEA projected that global nuclear capacity would be 2,100 gigawatts electric (GWe) in the year 2000. On the basis of resource information submitted by the major uranium-mining countries, it also estimated that the global resource of low-cost uranium was 3.5 million tons, enough to support only about 500 GWe of light-water reactor capacity for forty years.[12] Both projections were very far off.

At the end of 2013, global nuclear capacity was 373 GWe, and the IAEA projected that the global nuclear generating capacity in 2050 would be 440–1,113 GWe,[13] much less than its 1975 projection for the year 2000. And, while estimates of future nuclear capacity were being drastically downsized, estimates of global resources of low-cost uranium increased dramatically. In 2011, on the basis of national estimates of uranium resources, the Organisation for Economic Co-operation and Development (OECD) Nuclear Energy Agency and the IAEA projected a resource base of about 12 million tons of low-cost uranium, enough to support 2,400 GWe of nuclear capacity for forty years.[14] At the $130 per kilogram uranium (kgU) estimated recovery cost cutoff used by most countries to assess uranium reserves, the contribution of uranium to the cost of nuclear power would be 0.3 ¢/kilowatt hour (kWh), or only a few percent of the cost of electricity from a new nuclear power plant.[15] Geologists expect that, at higher recovery costs, far more uranium would become available.[16] This somewhat more costly uranium would have a minor impact on the cost of generating nuclear electricity.

The liquid-sodium-cooled breeder dream has not died everywhere, however. India and Russia are building demonstration breeder reactors and China is considering doing so. The nuclear establishments in these three countries project huge growth in their national nuclear generating capacities and are concerned about having to depend upon imported uranium. France also is designing a fast-neutron reactor, but for burning transuranic elements rather than breeding them.[17]

Plutonium Separation (Spent Fuel Reprocessing)

Based on the IAEA's 1975 nuclear growth projections, 1,400–2,000 tons of plutonium would have been required to provide startup cores for the 200 GWe fleet of breeder reactors expected by 2000.[18] Most of the countries with breeder development programs therefore launched programs to reprocess the spent fuel being discharged by their first-generation power reactors to recover plutonium for initial breeder reactor cores. The large reprocessing programs of France, Japan, and the United Kingdom were launched on this basis. All of these programs were driven by government policy.

In the United States during the 1960s and early 1970s, the government encouraged the building of privately owned reprocessing plants. Three were launched. One operated at West Valley, New York, for six years (1966–1972) but was then abandoned because of the need for costly upgrades to reduce worker radiation doses. It was handed over to the government and became a $5 billion federal-state cleanup project.[19] A second commercial reprocessing plant, built by General Electric in Morris, Illinois, was not put into operation because of a design flaw. Construction of a third, at Barnwell, South Carolina, was halted because of a change in U.S. policy following India's 1974 nuclear test.

The United States had provided India with technical assistance and training in reprocessing on the understanding that the technology would be used for peaceful purposes, specifically for India's breeder program. But India used the first plutonium that it separated for a "peaceful nuclear explosion." Jimmy Carter made U.S. encouragement of plutonium separation a political issue in the 1976 U.S. presidential election. Just before the election, President Ford agreed that "reprocessing and recycling of plutonium should not proceed unless there is sound reason to conclude that the world community can effectively overcome the associated risks of proliferation."[20] Soon after he came into office, President Carter announced that "we will defer indefinitely the commercial reprocessing and recycling of plutonium produced in the U.S. nuclear power programs [and the] plant at Barnwell, South Carolina, will receive neither federal encouragement nor funding for its completion as a reprocessing facility."[21] Carter succeeded in persuading the Nuclear Regulatory Commission to halt the process of licensing both the Barnwell plant and the Department of Energy's Clinch River Breeder Reactor demonstration project, but Congress kept both projects alive and, in 1981, Carter's successor Ronald Reagan announced that he was "lifting the indefinite ban which previous administrations placed on commercial reprocessing activities in the United States."[22]

President Reagan made clear, however, that the government would not subsidize reprocessing. This led the investors in the Barnwell

Reprocessing Plant to abandon the project. Congress then passed the Nuclear Waste Policy Act of 1982, which gave the Department of Energy the responsibility for constructing a deep geological repository for spent U.S. power reactor fuel to be funded by a fee of 0.1 ¢/kWh for all nuclear generated electricity.

Reagan also lifted the suspension on the licensing process for the Clinch River Breeder Reactor. The project had been launched in 1973 with equal funding commitments from utilities and Department of Energy, but with the DOE responsible for covering any cost increases. In 1983, after the estimated cost for the project had increased more than fivefold,[23] Congress canceled it.[24]

Thus, a combination of proliferation concerns and cost increases had ended the U.S. effort to commercialize plutonium as a fuel.

Most other countries that were already committed to reprocessing did not follow the U.S. lead, however. The United Kingdom and France continued their reprocessing programs. In 2013, India, which was building a prototype fast-neutron breeder reactor, had three small reprocessing plants and was building a larger one to provide plutonium for follow-on breeders.[25] Russia has been operating a pilot civilian reprocessing plant since 1977 and is building a second pilot plant. China also has a pilot-scale reprocessing plant and is considering building a larger one. Appendix 2 lists the major operational reprocessing plants.

Most countries with nuclear power have decided, however, to adopt interim storage instead of reprocessing their spent power reactor fuel; of the thirty countries (and Taiwan) that used nuclear power as of 2013, twenty-three countries did not reprocess their spent fuel. Some of these countries had previously chosen for a period to send their fuel abroad for reprocessing. Germany, for example, started to build a reprocessing plant in the 1980s but, when costs escalated and public opposition against the project grew, decided to contract with France and the United Kingdom for reprocessing services instead.[26] Belgium, Japan, the Netherlands, and Switzerland also became customers of France and the United Kingdom. Of these foreign customers, only the Netherlands, with one small nuclear power reactor, later renewed its contract. In 2012, the United Kingdom decided to end its reprocessing upon completion of its existing contracts, then projected for 2018.[27]

Similarly, Armenia, Bulgaria, the Czech Republic, Finland, and Slovakia sent spent fuel to Russia for reprocessing but chose to end these contracts. As of 2013, a small fraction of Ukraine's spent fuel continued to be shipped for reprocessing in Russia.

Separated Plutonium and International Security

Spent light water reactor (LWR) fuel contains about 1 percent pluto-nium.[28] Separating plutonium creates potential opportunities for nuclear weapon acquisition by both countries and subnational groups. The repro-cessing of 800 tons LWR fuel annually, the design throughput of Japan's Rokkasho Reprocessing Plant, for example, would yield about 8 tons of separated plutonium—enough for more than 1,000 nuclear weapons.

The national proliferation route through "civilian" reprocessing has played out repeatedly. Like India, North Korea characterized its reprocessing plant as civilian until it started to operate the plant. Other countries—including Argentina, Brazil, South Korea, Sweden, and Taiwan—acquired or sought to acquire reprocessing plants as a route to weapons but abandoned these efforts as a result of internal political change, pressure from the United States, or often both.[29]

Even when a country joins the Non-Proliferation Treaty and accepts IAEA safeguards, the quantities of plutonium being separated at a repro-cessing plant can be so large that measurement uncertainties could allow the slow diversion of sufficient plutonium to make many nuclear weapons without detection. Japan's Tokai Pilot Reprocessing Plant, for example, reported in January 2003 that it had cumulatively recovered 6,900 kilo-grams of plutonium but had a discrepancy of 206 kilograms. Later it found that 106 kilograms of the missing plutonium had gone into high-level waste (HLW), calculated that 29 kilograms of plutonium-241 had decayed, and speculated that 12 kilograms had been disposed of with transuranic waste. This left a discrepancy of 59 kilograms, about one percent of the plant's throughput, which is typical of input-output mea-surement differences in reprocessing plants and also enough for several nuclear explosives.[30]

Reprocessing also makes plutonium accessible to subnational groups that otherwise would find it virtually impossible to obtain for either a nuclear weapon or a radiological dispersal device.[31] In spent nuclear fuel, plutonium is protected by the gamma radiation field that is generated by fission products—especially by cesium-137, which has a half-life of thirty years. Even after fifty years, an unshielded person standing a meter away from a pressurized water reactor fuel assembly would accumulate a life-threatening dose within less than an hour.[32] Separated plutonium does not have a significant gamma radiation field, however, and therefore can be safely handled in a sealed package and could be converted into the pit of a nuclear explosive in a glove box (figure 6.2).[33] After a century

Figure 6.2

A worker at Russia's Mayak Reprocessing Plant in the Urals in 1994 packing a container holding 2.5 kilograms of plutonium in the form of plutonium oxide powder into an outer container. The dose rate of penetrating gamma radiation from the containers is small but a milligram of plutonium oxide inhaled collectively by a large group of people probabilistically would cause about ten cancer deaths.
Source: Authors' archives.

or so, the gamma radiation field in spent fuel will have largely decayed away, at which point protection against access to the plutonium it contains will depend more on monitoring, physical security measures, and, possibly, deep burial.

Plutonium Recycle in Light Water Reactors

When breeder reactors failed to be economically competitive, countries that were separating plutonium had to decide what to do with their accumulating stockpiles. An obvious strategy was to recycle the plutonium back into fuel for the light water reactors (LWRs) that produced it. LWRs are ordinarily fueled with low-enriched uranium (LEU) containing 4–5 percent uranium-235. A mixture of 6–8 percent reactor-grade plutonium with depleted uranium has about the same fuel value.[34] Unless the control system of a reactor has been designed for MOX fuel; however, the fraction of MOX fuel that can be used in a reactor core is limited to typically less than one third.

Given that the amount of plutonium in spent LEU fuel is about 1 percent and the percentage in fresh MOX fuel is 6–8 percent, the plutonium from 6–8 tons of spent LEU fuel is required to make a ton of MOX fuel to replace 1 ton of LEU fuel. Plutonium recycle in LWRs therefore reduces their uranium requirements by about 15 percent. Multiple recycles would increase this percentage but has not yet been done because once-recycled plutonium contains a reduced percentage of the isotopes plutonium-239 and plutonium-241 that chain-react in LWRs.[35] Re-enriching and recycling the recovered uranium can realize additional uranium savings of about 10 percent.[36]

France has led in plutonium and uranium recycle in LWRs. In 2000, however, a study of the economics of reprocessing commissioned by French Prime Minister Lionel Jospin found that reprocessing was increasing the cost of nuclear of nuclear power in France by the equivalent of about 0.2 ¢/kWh or about $34 billion (2006 $) over the forty-five-year average lifetime of France's nuclear power reactors. Even considering the investment in the reprocessing plant a sunk cost, the report found that France could save $8 billion if it abandoned reprocessing in 2010.[37]

With the end of significant reprocessing of foreign spent fuel in France, the full economic burden of supporting the French reprocessing complex falls on the government-owned national utility, Électricité de France (EDF). EDF has struggled with the nuclear services company AREVA—also government-owned—over the price of reprocessing.

Reprocessing is currently required by law in France, however, and as of 2013 no government had yet come to power in France that was willing to take on the challenge of shutting down AREVA's huge La Hague reprocessing plant with its six thousand employees in a rural area of France.

The use of plutonium as MOX reactor fuel brings with it a set of security risks associated with storage, processing, and transport. As of the end of 2012, about 245 tons of the world's separated civilian plutonium, more than 90 percent of the total, were stored at four sites in Europe and Russia. These were the French reprocessing and fuel fabrication sites at La Hague and Marcoule (together about 80 tons, as of December 2012), the British site at Sellafield (117.3 tons), and Russia's Mayak facility (49.2 tons as of the end of 2012). From a security perspective, the consolidation of separated plutonium at the fewest number of possible sites is to be preferred if these sites are intended as secure long-term storage. The material becomes more vulnerable to theft and dispersal when it is in transit to be fabricated into MOX and then during delivery to reactors as fuel.

Due to the use as MOX of a large fraction of plutonium that is separated in France, there were in 2012 frequent transports of large quantities of separated plutonium over long distances across the country and as MOX to some other Western European countries. In France, each year an average of 200 kilograms of plutonium was transported across the country in each of about sixty shipments from the reprocessing plant at La Hague to Marcoule for MOX fuel fabrication.[38] The MOX fuel, containing 10–13 tons of separated plutonium, was transported each year over distances up to 1,000 kilometers to supply domestic and foreign reactors. MOX-fueled reactors each need about 500 kilograms of plutonium per year, which may be delivered in one or several shipments. Thus, during an average week, at least one shipment of a few hundred kilograms of plutonium—enough to make anywhere from thirty to sixty Nagasaki bombs—was on the roads in France.

MOX fuel was also sent by sea in armed ships, most notably from Europe to Japan. In 2012, there were about 35 tons of Japanese separated plutonium stored in France and the United Kingdom awaiting shipment to Japan in the form of MOX fuel. As of the end of 2013, Japan had received a total of five MOX shipments from Europe since 1999, containing a total of 4.5 tons of plutonium, of which 2.5 tons had been loaded into reactors. These shipments raised safety and security concerns in the United States, in Japan, and in countries along each of the three ocean routes that had been taken.[39]

The problem of transporting plutonium over large distances would become more acute if there were increased international reliance on plutonium separation and use.

Reprocessing and Radioactive Waste Disposal

Spent fuel contains dangerous fission products and transuranic elements, the latter including long-lived isotopes of neptunium, plutonium, and americium. There is a general consensus in the nuclear energy community that, if spent fuel is not reprocessed, it should be buried underground (at least 300 meters deep) in corrosion-resistant containers within a carefully selected geological medium to minimize the probability of long-lived isotopes reaching the surface.

The political and technical challenges confronting efforts at long-term storage and disposal of spent fuel from nuclear power reactors and high-level radioactive waste from reprocessing operations have thus far prevented the licensing of a geological repository for spent fuel or high-level reprocessing waste anywhere in the world.[40] In particular, finding sites for repositories has proven politically very difficult. Almost all countries that have tried to site repositories have had one or more failures; a notable example is the plan for a repository under Yucca Mountain in Nevada in the United States, which was finally abandoned in 2010 after more than twenty years of studies to support a site license application at a cost of $15 billion.

It appears that voluntary and consultative processes for siting geological repositories have been more successful than top-down decision making. Finland and Sweden, which have developed a participatory approach, have both had success in obtaining acceptance of a repository site. In each case, that site is adjacent to a nuclear power plant. Spent nuclear fuel disposal plans there may, however, face technical problems related to the long-term durability against corrosion of the copper canisters intended to contain the spent fuel. No country appears ready to host a multinational spent fuel facility, which will face the same siting and licensing issues that confront national repository efforts and possibly more public opposition. Given the difficulties of repository siting, there is a need for transitional measures to manage spent fuel and high-level waste for periods of decades and possibly a hundred or more years.

For the majority of countries that do not reprocess, when a reactor's spent fuel storage pool is full, the oldest spent fuel is put into interim storage in air-cooled dry casks on the nuclear power plant site to make space in the pool for additional discharges.

Advocates of reprocessing argue that, quite apart from the breeder dream, reprocessing reduces the risks from spent fuel disposal by removing long-lived plutonium and could be used to remove other long-lived transuranic isotopes from the high-level waste. The elimination of the separated transuranics would require their fissioning, however, and fast neutron reactors can fission transuranic isotopes that cannot be effectively fissioned by the slow neutrons in light water reactors. Thus, advocacy of sodium-cooled plutonium breeder reactors has morphed into advocacy of sodium-cooled reactors with redesigned cores for net "burning" (fissioning) of transuranics.

In 1992, the U.S. Department of Energy asked the U.S. National Academy of Sciences (NAS) to examine the costs and benefits of eliminating the long-lived transuranic elements in spent LWR fuel. The NAS committee concluded that the cost would be much higher than that of storage and direct burial. It also found that, in deep geological environments where the water has been trapped long enough so that its oxygen has been consumed by reactions with the rock, the transuranics would be insoluble and therefore not mobile in ground water.[41] It concluded, therefore, that "none of the dose reductions seem large enough to warrant the expense and additional operational risk of transmutation."[42] The committee included in "operational risks" radiation doses to workers in the reprocessing and fuel fabrication facilities and the public. These dose increases would tend to offset reductions in doses millennia in the future from any transuranic leakage from deep underground repositories.

Without separation and transmutation, after one hundred years or so the plutonium in the spent fuel will no longer be protected by the gamma radiation from fission products with which it is mixed. Spent fuel in sealed 100-ton casks is difficult to steal, however, and burial in a deep geological repository would make access even more difficult. Advocates of reprocessing worry that, in the distant future, such repositories could become "plutonium mines." But this future danger has to be balanced against danger of diversion from the "plutonium rivers" that reprocessing creates. Reprocessing and recycle variants have been proposed in which plutonium would not be separated from other transuranics or even some relative short-lived fission products. It has been found, however, that the reprocessing facilities could quickly be modified to produce pure plutonium.[43]

Reprocessing advocates argue that, with the uranium removed, reprocessing reduces the volume of the waste that must be disposed of in a deep repository. But this refers only to the high-level (i.e., concentrated) waste. When all reprocessing wastes that require deep burial are included, there is no longer a clear volume reduction from reprocessing.[44]

Indeed, reprocessing complicates the radioactive waste disposal problem by turning one waste form, spent fuel, into multiple waste forms:

• HLW immobilized in glass in a "vitrification" process;[45]

• plutonium-containing wastes from the MOX fuel fabrication process;

• spent MOX fuel, since, in the absence of the commercialization of breeder reactors, it appears unlikely that MOX fuel will be reprocessed and the contained plutonium recycled again;[46] and

• additional wastes, including ultimately the contaminated parts of the reprocessing and MOX fuel fabrication plants themselves.

Also, the higher the heat output of radioactive waste, the more it has to be spaced out underground to avoid overheating the rock. The radioactive heat output of the waste therefore is a more important determinant of the required area of a repository than the volume of the waste. Because of the high heat output of spent MOX fuel, here again there is little repository benefit from reprocessing with one recycle of plutonium in MOX fuel as is currently practiced in France.[47]

Finally, cleaning up a reprocessing site is a costly business. In Japan, the 2011 estimate for decommissioning the Rokkasho Reprocessing Plant was ¥1.9 trillion (~$20 billion). In the United Kingdom, the 2013 estimate for decommissioning the Sellafield reprocessing site was £67.5 billion (~$110 billion).[48]

Some foreign utilities sent their spent fuel to France and the United Kingdom to be reprocessed as a way to deal with the domestic criticism that they did not have a path forward for disposing of their spent fuel. But they exported their political problems only temporarily. France and the United Kingdom insisted that they would not become, as one British campaigner put it, the "the nuclear dustbin of the world."[49] The solidified concentrated radioactive waste from the reprocessing of foreign spent fuel therefore is being sent back to the customers. Since managing the returning reprocessing waste is no less problematic than managing the original spent fuel, all but the Netherlands did not renew their

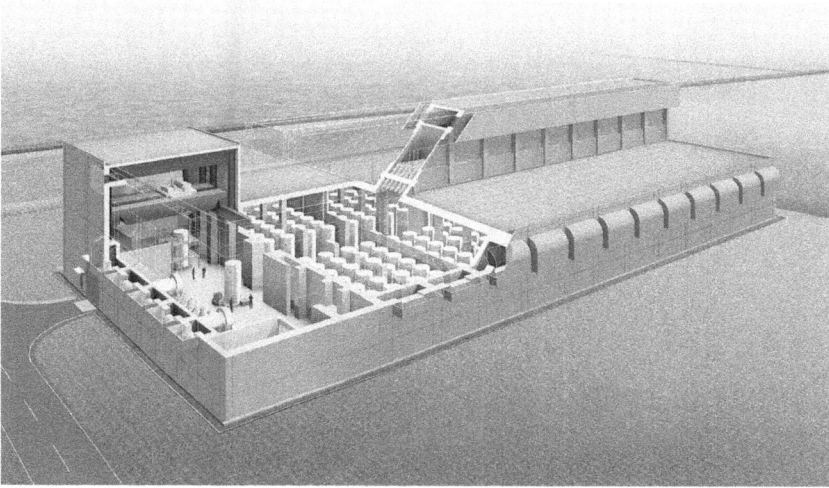

Figure 6.3

Artist's conception of a 3,000-ton-capacity interim spent fuel storage facility near Mutsu, Japan, that was completed in 2013. Each cask will store about 10 tons of spent fuel. The radioactive decay heat generated by the spent fuel will be removed from the cask surfaces by air that is circulated by natural convection facilitated by the stack on the top of the protective building. In the United States, spent fuel storage casks are stored outside.
Source: Recyclable-Fuel Storage Company.

reprocessing contracts. The others, with the exception of Japan, began to build interim air-cooled dry cask surface storage for their spent fuel pending the availability of deep underground repositories.

Japan's utilities believed that it would be politically impossible to expand spent fuel storage at their nuclear power plants. They therefore built their own reprocessing plant and are using the same site to store the solidified high-level reprocessing waste coming back from Europe. As of 2013, however, commercial operation of Japan's Rokkasho Reprocessing Plant was over fifteen years overdue and Japan was building interim dry-cask spent fuel storage capacity nearby (figure 6.3). In 2011, Japan's Atomic Energy Commission concluded that reprocessing will cost Japan ¥0.8–1.0/kWh (~1 ¢/kWh) more than direct disposal.[50] Over the design lifetime of the reprocessing plant, that will add up to about ¥10 trillion (~$100 billion).[51]

Table 6.2
Global stocks of separated civilian plutonium as of the end of 2012

	Separated civilian plutonium in country (metric tons)	Of which foreign-owned (metric tons)
China	0.01	0
France	80.6	22.2
India	4.9 ± 0.3	0
Japan	9.3	0
Russia	50.7	0.003
United Kingdom	120.2	23.8
Totals	265.8	46.0

Source: Aside from India, the other five countries report these numbers annually to the IAEA under INFCIRC/549, *Communication Received from Certain Member States Concerning Their Policies Regarding the Management of Plutonium.* The estimate for India is from *Global Fissile Material Report 2013,* chap. 1. Although this material was intended to fuel India's Prototype Fast Breeder Reactor, a civilian use, India has declared that it is not available for international safeguards. India may wish to reserve the option to use weapon-grade plutonium produced in the PFBR blanket for weapons.

Civilian Plutonium Stocks

The failure of breeder reactor commercialization programs and technical and political delays in the recycle of plutonium into light water reactor fuel have resulted in global stocks of separated civilian plutonium of about 260 tons as of the end of 2012—more than the Soviet Union and United States produced for nuclear weapons during the Cold War (table 6.2).[52] Although little of the civilian plutonium is weapon-grade (i.e., contains more than 90 percent plutonium-239), all is weapon-usable as discussed in chapter 2.

As of 2012, France was recycling its separated plutonium in light water reactor fuel while separating more, not maintaining an approximately steady-state situation. Japan was planning to recycle also but its reprocessing plant has been delayed by technical breakdowns and its plutonium recycling program has been delayed by public opposition.

The United Kingdom completed a MOX fuel fabrication plant in 2001 to recycle the plutonium it was separating for its foreign customers—especially Japan—but abandoned that plant in 2011 after technical problems limited its output during its first ten years to only one percent of its design capacity.[53] A few months later, the UK government proposed to build another MOX plant to recycle the United Kingdom's own stockpile of about 100 tons of separated plutonium if new light water reactors are built in the United Kingdom to replace its aging gas-cooled reactors.[54]

China, India, and Russia currently plan to use their separated plutonium for initial cores for breeder reactors. Other options for plutonium disposition are discussed in chapter 9.

The greatest danger from reprocessing is that it opens a route to weapons for nations—and potentially for terrorists as well. Based on experience to date, despite its potential benefit of making nuclear power more uranium efficient, reprocessing will not have a net economic benefit for the foreseeable future. Additional plutonium separation can be postponed indefinitely while the world struggles to resolve the problem of safely disposing of its huge legacy of already separated plutonium.

7

Ending the Use of HEU as a Reactor Fuel

As of 2013, virtually all nuclear power reactors are fueled by natural or low-enriched uranium enriched to less than 5 percent in uranium-235. LEU is not weapon-usable. There are hundreds of compact research reactors and naval propulsion reactors, however, that use highly enriched uranium as fuel. This HEU is a potential source of nuclear weapons material for governments interested in acquiring nuclear weapons and for would-be nuclear terrorists. Whether a reactor is fueled with natural or low-enriched uranium or HEU therefore has important implications with regard to the dangers of nuclear proliferation and nuclear terrorism and for progress in reducing and eventually eliminating global fissile material stockpiles.

The first nuclear reactors—starting with Enrico Fermi's critical assembly—were fueled with natural uranium in which only one out of 140 nuclei is uranium-235 (see chapter 2, this volume). Natural uranium cannot sustain a chain reaction with the fast neutrons released by fission. Too many neutrons would be absorbed by uranium-238 nuclei without fissioning. Neutrons can be slowed down, however, using a moderator material to speeds at which they are preferentially absorbed by uranium-235 and cause fissions, so that a slow, non-explosive chain reaction can be sustained even in natural uranium.

The moderation requirement of natural-uranium-fueled reactors results in large spacing between fuel elements and therefore large reactor cores, however. The demand for nuclear reactors for submarines and for research reactors with high neutron fluxes provided an incentive for the development of reactors with more compact cores. In the United States and Soviet Union, the availability of large quantities of weapon-grade uranium made it a natural choice for such applications.

By the 1960s, "weapon-grade" uranium (93.5 percent uranium-235 in the United States) had become the standard fuel for U.S.-designed research reactors. Eighty to ninety percent enriched HEU was similarly standard in Soviet-designed research reactors.[1] HEU fuel also was adopted for naval propulsion by the United States, the Soviet Union, United Kingdom, France, and most recently by India.

In the 1970s, amid growing concern about proliferation, the United States and Soviet Union began an effort to roll back the use of HEU fuel for research reactors that they had exported around the world. These concerns accelerated after the events of September 11, 2001, when the United States also became seriously concerned about the possibility that, if terrorists were able to penetrate an HEU storage facility, they might be able to put together and detonate an "improvised nuclear explosive" quickly. In an October 2000 force-on-force test of the security at the Los Alamos Critical Experiments Facility, several "mock terrorists" penetrated the facility while large plates of HEU were outside its vault. The protective force was unable to drive the intruders out and, as a result, the attackers might have had time to create an improvised nuclear explosive with the plates.[2]

The first Nuclear Security Summit, which brought the leaders of forty-seven countries together in Washington, DC, in 2010, sought to bring global attention to the imperatives of reducing the use of HEU as a fuel, minimizing the number of locations where HEU can be found, and increasing security for HEU storage and transport. In the Final Declaration, the states agreed that "highly enriched uranium and separated plutonium require special precautions and [we] agree to promote measures to secure, account for, and consolidate these materials, as appropriate; and encourage the conversion of reactors from highly enriched to low enriched uranium fuel and minimization of use of highly enriched uranium, where technically and economically feasible."[3]

Most attention has focused on ending the use of HEU in civilian research reactors, which are seen as particularly vulnerable. But HEU in naval fuel cycles also is a potential theft target during fabrication, storage, and transport. Indeed, one of the incidents that helped galvanize the U.S. effort to help Russia secure its nuclear materials after the collapse of the Soviet Union was a theft in 1993 of HEU submarine fuel from a storage facility in Murmansk. The special investigator concluded, "Potatoes were guarded better."[4] Similarly, in the United States, it appears that, with the collusion of plant management, hundreds of kilograms of weapon-grade uranium were diverted in the mid-1960s from a U.S. naval fuel fabrication facility and shipped to Israel for use in its weapons program.[5]

As of 2013, about 2 tons of weapon-grade uranium—enough for about forty gun-type weapons or 100–150 implosion explosives—flowed through the U.S. naval fuel cycle every year.[6] The huge quantities of HEU that have been placed in storage for future use in naval reactor fuel also could make future negotiated reductions in the nuclear weapon stockpiles more difficult. The United States has designated 152 tons of excess Cold War weapon-grade uranium for this purpose,[7] enough for 5,000–10,000 nuclear warheads. Thus there are reasons to consider designing future naval reactors to use LEU fuel.

This chapter looks at the use of HEU in reactor fuel and the progress and challenges of ending this application for both research and naval propulsion reactors.

Research Reactors

Research reactors can be divided into three categories: steady-power reactors; critical assemblies, operated at very low power to check calculations of the neutronics of proposed core designs; and pulsed reactors, mostly used to simulate the effects of neutron bursts from nearby nuclear explosions on electronics in satellites or warheads in space.

Steady-power research reactors are used primarily as sources of neutrons to irradiate materials for various purposes including: testing the effects of neutron radiation on candidate reactor structural materials and fuels; producing radioactive isotopes for diagnostic medical tests; and probing the structures of crystals of complex materials. The more intense the neutron flux from a reactor, the shorter the necessary irradiation time and the more productive the facility.

For a given core radius, the neutron flux increases with reactor power; and for a given power, the flux at the core surface increases as the core radius is reduced. Neutron flux also rises if neutron absorbers within the core are reduced. Uranium-238 is the dispensable neutron absorber in uranium because most of the neutrons it absorbs do not cause fissions. With HEU fuel, neutron absorption by uranium-238 is largely eliminated and, without uranium-238 to dilute it, the amount of uranium-235 that can be packed into a given volume of fuel is maximized. For example, the Oak Ridge National Laboratory's 85 MWt High Flux Isotope Reactor (HFIR), which came online in 1961 and was still operating in 2013, had a core volume of only 0.1 cubic meters—about the size of an automobile engine.

Figure 7.1

MTR-type Research Reactors. The original Material Testing Reactor was designed and built collaboratively by the Oak Ridge and Argonne National Laboratories and operated from 1952–1970 on what is now the site of the Idaho National Engineering Laboratory. Alvin Weinberg, then the Director of Oak Ridge National Laboratory, presented the basic design of this reactor in great detail at the Atoms for Peace Conference held in August 1955 in Geneva.
Source: Idaho National Laboratory.

Figure 7.1 shows one of the first research reactors, the 30 megawatts thermal (MWt) Material Testing Reactor. Its fuel design, using highly enriched uranium dispersed in thin aluminum-clad plates, became the de facto standard for most research reactors built since.

After President Eisenhower's "Atoms for Peace Speech" at the United Nations in 1953 and the 1955 Atoms for Peace Conference in Geneva, research reactors quickly spread worldwide—almost all provided by the United States or Soviet Union. The number of countries with research reactors increased from five in 1955 (the United States, Canada, Soviet Union, the United Kingdom, and France) to thirty-one in 1960 and forty-eight in 1965, including twenty-nine developing countries.[8] Initially, the United States supplied LEU fuel to foreign users.[9] In the early 1960s,

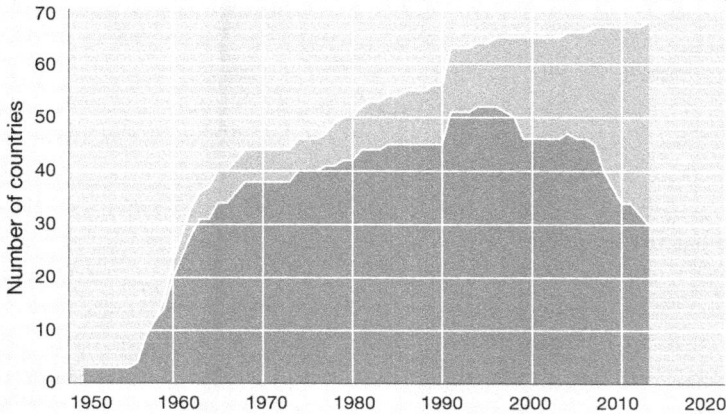

Figure 7.2

Countries with HEU-fueled and LEU-fueled research reactors. After President Eisenhower's 1953 "Atoms for Peace" Speech, the United States and Soviet Union exported HEU-fueled research reactors to about 40 countries (number of countries with HEU fuel shown in dark gray). In 1978, both the United States and Soviet Union began to encourage conversion to low-enriched uranium but the programs were quite small (number of countries with only LEU-fueled research reactors shown in light gray). Since 2002, the United States has greatly expanded funding for these efforts. The dates shown are when a country began operating a research reactor that was then or later HEU-fueled. Some of these reactors were fueled for their first few years with low-enriched uranium. A country is considered to be cleared of HEU when the amount of residual HEU in the country is less than 1 kilogram.

however, as U.S. HEU production peaked, the United States began to export annually hundreds of kilograms of HEU in research reactor fuel. Between 1965 and 1971, the United States exported 13 tons of HEU for such use—enough for about 250 Hiroshima bombs.[10] The Soviet Union also shifted toward HEU.[11] By 1965, of the forty-eight countries with research reactors, at least thirty-four had research reactors fueled with HEU (figure 7.2).[12]

Converting HEU-Fueled Research Reactors to LEU

In 1974, India used its first separated plutonium, which had been produced and separated with U.S. Atoms for Peace assistance, to test a nuclear weapon design. This led the United States to adopt a policy of discouraging the use of weapon-usable materials, specifically, HEU and plutonium, as civilian reactor fuels. During 1977 and 1978, at U.S. urging, a yearlong International Nuclear Fuel Cycle Evaluation (INFCE) was held at IAEA headquarters in Vienna to discuss the trade-offs involved. One of its findings was the following:[13] "A change to lower enrichment (either below 20 percent or to around 45 percent) would seem to be feasible for the great majority of [research] reactors." Both the United States and the Soviet Union launched Reduced Enrichment for Research and Test Reactor programs in 1978 that focused initially on the fuels that they were shipping abroad.[14] In 1986, however, the U.S. Nuclear Regulatory Commission (NRC) broadened this effort by requiring conversion of U.S. domestic civilian research reactors, conditioned on the availability of lower-enriched fuel and government funding for conversion-related costs.[15] The cost of conversion of an HEU-fueled research reactor typically is of the order of one million dollars.

The United States chose for the converted reactors a reduced enrichment target of 19.75 percent, just below the 20 percent boundary between low-enriched and highly enriched uranium. Russia initially chose an intermediate target of 36 percent but, in 1993, after the United States began to fund conversion of Soviet-designed research reactors in third countries, adopted the 20 percent target as well.[16] Other countries that fabricated research reactor fuel also joined in the LEU fuel development effort, including Argentina, Canada, France, Germany, and South Korea.[17]

Unlike the United States, Russia did not make conversion of its domestic research reactors an early priority. In late 2012, however, ROSATOM, Russia's State Atomic Energy Corporation, announced that it had developed a program to convert different kinds of research reactors in Russia starting in 2013.[18] Conversion of one reactor, and possibly a second, was expected by the end of 2014 and further U.S-Russian cooperation on reactor conversion was expected.[19] This initiative followed a two-year joint U.S.-Russian conversion feasibility study of six Russian research reactors that included five of Russia's fifteen operating steady-state research reactors as well as a pulsed reactor. In addition, by 2013, nine of Russia's HEU-fueled research reactors had been shut down and were either being or had been decommissioned.[20]

In many cases, reactor operators have been reluctant to convert, and their acceptance of conversion has required analytical efforts to determine the impact of conversion on reactor performance on a reactor-by-reactor basis. In 1992, to help overcome resistance by foreign research reactor operators to even the consideration of conversion, the U.S. Congress passed the "Schumer Amendment" to the Atomic Energy Act.[21] The amendment instructs the U.S. Nuclear Regulatory Commission to issue a license for the export of HEU fuel to a foreign research reactor *only* if no LEU fuel is available for that reactor and then *only* if LEU fuel is under development and the reactor operator has committed to convert as soon as the new fuel becomes available. Russia appears to have adopted a similar policy of not exporting HEU fuel if alternative LEU fuel is available.

To support the reactor conversion effort, programs have been designed to replace HEU with LEU fuel in a way "that shall be practically invisible for operators and users of experimental facilities."[22] Specifically, conversion is generally accomplished while keeping the same fuel geometry and approximately the same loading of uranium-235, typically kilogram quantities. To dilute weapon-grade uranium with uranium-238 down to an enrichment level of less than 20 percent without changing the geometry, the uranium density in the fuel must be increased roughly fivefold.[23] Unless the reactor power can be increased, however, the neutron flux available for experiments after conversion is typically reduced by 5 to 15 percent because the added uranium-238 in the fuel absorbs some of the neutrons released by fission. Usually, however, improvements can be made to the neutron beams and experimental equipment to more than offset this loss.[24]

Some reactors require the development of fuels with very high uranium density before they can be converted. The development of such fuels has focused on a high-density uranium metal alloy containing up to 10 weight-percent molybdenum.[25] Unfortunately, the introduction of these fuels has been delayed by technical problems. If these problems can be overcome, it should be possible to end almost completely the use of HEU fuel in research reactors.

As of April 2013, sixty-five research reactors had been converted to LEU, including eighteen in the United States.[26] Some of the reactors that have been converted have subsequently been retired. Between 1978 and 2010, 127 old HEU-fueled research reactors were retired, about twice the number converted.[27] In the meantime, two new HEU-fueled research reactors had been built: Germany's 20-MWt FRM-II commenced operation in 2004 and Russia's PIK reactor, near Leningrad, with a power of 100 MWt was completed, although still not in operation in 2013. France's

100 MWt Jules Horowitz Reactor (JHR) was expected to start up with HEU fuel because of delays in the development of high-density LEU fuel. These three reactors would together require 150 to 200 kilograms of weapon-grade HEU fuel per year.

As of late 2013, there were still about forty-five steady-state HEU-fueled research reactors operating globally (table 7.1). Ten were very small Canadian and Chinese-designed reactors whose lifetime cores contain about 1 kilogram of weapon-grade uranium and therefore are not of great proliferation concern—but some of those were being converted.[28] Of the remainder, it was believed twenty-seven could be converted using the higher density fuels that were still under development.[29]

In the longer term, there is a growing expectation that other types of neutron sources could take on some of the missions currently carried out using research reactors. Most important, accelerator-driven spallation neutron sources such as the Spallation Neutron Source at the U.S. Oak Ridge National Laboratory and the planned European Spallation Source in Sweden provide intense pulsed neutron fluxes for research with neutrons and have important advantages compared to steady-state fission reactors.

In addition, low-power aqueous solution reactors, fueled with low-enriched fuel, have been proposed as an alternative to the use of high-power reactors for molybdenum-99 production.[30] Molybdenum-99 production is the main commercial application of today's research reactors; its daughter product, technetium-99m, is the most commonly used medical radioisotope and is used in about 30 million diagnostic procedures every year.

As table 7.1 shows, however, steady-state research reactors are only a part of the universe of HEU-fueled research reactors. As of late-2013, HEU fuel was being used in close to forty critical and subcritical assemblies, over twenty pulsed reactors, two dedicated isotope-production reactors, two breeder reactors, and about 200 naval and icebreaker propulsion reactors.

Critical assemblies. Critical assemblies are mockups of reactor cores that are used principally to check computer calculations of the criticality of proposed core designs. They are simple and can be built up and torn down easily because they do not have cooling systems. The power generated is typically in the kilowatt range and not sustained, and heat is removed by natural air circulation. As civilian research reactors convert to LEU, the rationale of many HEU-fueled critical assemblies

Table 7.1
HEU-fueled reactors worldwide, by region and reactor type, late 2013

	Critical and subcritical assemblies	Pulsed reactors	Research reactors		Isotope production reactors	Breeder reactors	Naval reactors	Total
			<0.25 MWt	0.25–250 MWt				
Russia	23	15	2	13	2	1	84	140
China	1	–	2	0		1	0	4
Europe	3	3	3	5			13 (United Kingdom)	27
USA	6	2	1	6			103	118
Others	5	2	9	4			2 (India)	22
Total	38	22	17	28	2	2	202	311

Source: "Facilities: Research and Isotope Production Reactors," http://fissilematerials.org. The list of naval reactors includes military vessels, icebreakers, and land-based naval training reactors. For naval reactors, see Stephen Saunders, *IHS Jane's Fighting Ships 2012–2013* (Coulsdon, Surrey: IHS Jane's, 2012).

will disappear. That will still leave critical assemblies that are mockups of HEU-fueled naval reactors and fast-neutron reactors, however (figure 7.3a).

Fortunately, computers are so powerful today that they can simulate reactors in great detail and be tested against hundreds of well-documented critical experiments.[31] In 2011, the United Kingdom decided to shut down its two land-based naval test reactors, in part because computer modeling has become so accurate.[32] In principle, only a few flexible critical assemblies are needed worldwide in case different or more precise benchmark experiments are required.[33]

In the United States, the upgrades in security and associated huge cost increases for protecting HEU at government laboratories since September 11, 2001, have provided an incentive to shut down unneeded HEU-fueled critical facilities. Four of the five remaining U.S. HEU critical assemblies have been moved from a location in Los Alamos that was not considered adequately secure to the high-security Device Assembly Facility on the Nevada Test Site.[34] In 2013, Russia accounted for more than half of all HEU-fueled critical assemblies worldwide (table 7.1). The incentives to shut down appear to be weaker in Russia, This may be partly because different ministries pay for research and for security.[35]

Pulsed reactors. The principal use of most pulsed reactors has been to study the potential effects on warheads and satellites in space of intense bursts of neutrons from nearby nuclear explosions. The fuel of a pulsed reactor is typically a hollow cylinder of weapon-grade uranium metal alloy (figure 7.3b). When neutron-absorbing control rods are withdrawn, the reactor goes supercritical and emits a pulse of fission neutrons to irradiate test objects inside its central cavity. The heat released by the fissions raises the temperature of the fuel, which expands, allowing a larger fraction of the neutrons to escape. As a result, the reactor goes subcritical within about a millisecond. The control rods are then reinserted to keep the reactor subcritical as it cools.

Pulsed reactor cores can contain hundreds of kilograms of barely irradiated weapon-grade HEU. As of 2013, Russia's fifteen HEU-pulsed reactors were two-thirds of the global total (table 7.1). The United States used to have a similar number but has retired and decommissioned all but two because of the high cost of security for the HEU.[36] In 2013, the U.S. Army, owner of the last operating U.S. pulsed reactor,

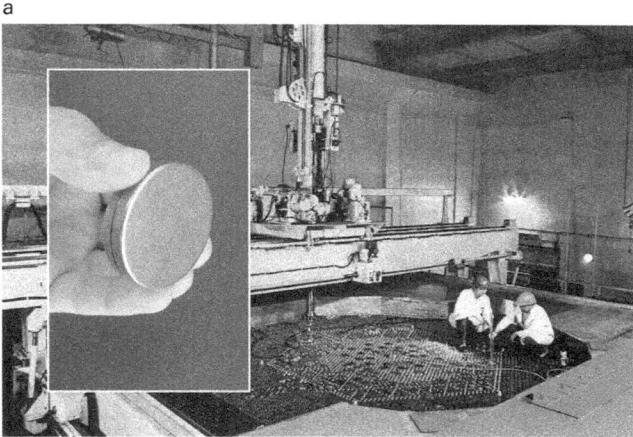

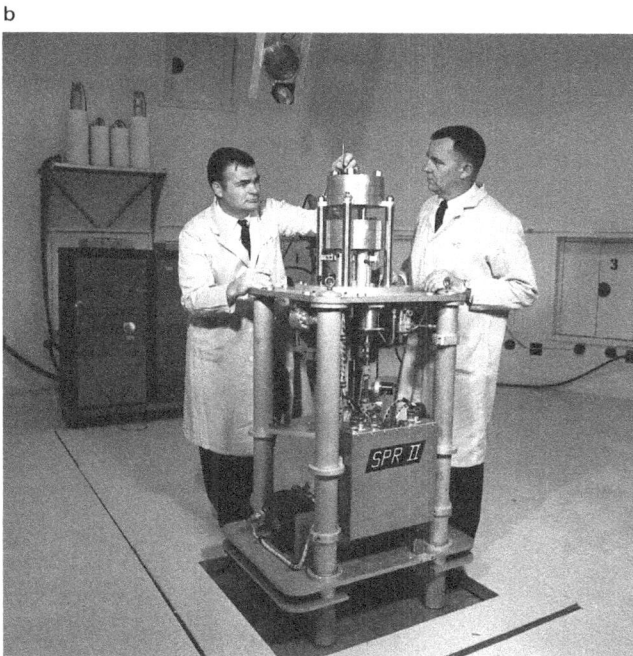

Figure 7.3

Critical assemblies and pulsed reactors often contain large amounts of HEU. (a): In the BFS-2 fast critical assembly at the Institute of Physics and Power Engineering in Obninsk, Russia, disks of plutonium, HEU, and depleted uranium are stacked up to simulate breeder-reactor cores with different enrichments of uranium and plutonium. The facility has an inventory of tens of thousands of disks containing a total of 3.5 tons of HEU and about 0.5 tons of plutonium. The inset shows one of these disks. (b): Pulsed reactors, such as the former Sandia Pulsed Reactor II (SPR II) at Sandia National Laboratory in Albuquerque, New Mexico, were used to create intense bursts of neutrons to verify that weapon parts and satellites could withstand high doses of radiation. SPR II contained more than 130 kilograms of weapon-grade uranium metal.

Source: U.S. Department of Energy.

awarded a contract to develop "neutron generator technologies that do not utilize highly enriched fissile material" that can be used to "meet the neutron and gamma flux requirements specified by the Army for nuclear survivability testing . . . and will greatly reduce the cost and security burden to the Army."[37] One alternative that was being pursued by the U.S. Sandia National Laboratory, which retired its own pulsed reactor in 2006, was to model on computers the effects of neutron pulses on electronics, starting at the level of the individual transistor.[38] Sandia also has designed an LEU-fueled pulsed subcritical reactor that would be driven by neutron pulses produced by an external nonreactor source.[39] In any case, here again, primarily because of the cost of guarding HEU, the United States was on a path to phasing out a whole class of HEU-fueled reactors. The challenge is to get Russia to focus on doing so as well.

Naval Propulsion Reactors

Along with a variety of research reactors, the weapon states also developed reactors for naval propulsion. The first land-based prototype submarine reactor began operating on the U.S. Idaho Test Reactor Site in 1953.[40] The United States launched *Nautilus,* the world's first nuclear-powered submarine, the following year, and the first nuclear powered aircraft carrier, the *Enterprise,* in 1961 (figure 7.4). The next four nuclear weapon states acquired nuclear submarines in the order in which they had acquired nuclear weapons: the Soviet Union (1958), the United Kingdom (1960), France (1967), and China (1974). By the late 1980s, just before the end of the Cold War, there were about four hundred nuclear-powered submarines and surface vessels deployed by these five countries with the vast majority operated by the Soviet Union and United States.[41]

After the end of the Cold War, the U.S. and Russian nuclear fleets were downsized substantially. In 2012, the United States had 79 nuclear-powered submarines and ships, Russia 44, the United Kingdom 11, France 7, China 9, and India 1 for a total of 155.

Given the worldwide deployment of its navy, the design priority of the United States has been to maximize the core life of its reactors. U.S. nuclear submarines and aircraft carriers are being built with "lifetime" cores, designed to last thirty to forty years. The United Kingdom uses U.S. fuel designs in its submarines and has migrated to lifetime cores as well. All U.S. naval-propulsion reactors were initially fueled with

a

b

Figure 7.4

The USS *Nautilus* (a) was the first nuclear-powered submarine and the USS *Enterprise* (b) was the first nuclear-powered aircraft carrier. The development of a nuclear reactor for submarine propulsion started in 1947 and construction of the *Nautilus* was authorized in 1951 and completed in 1954. It was decommissioned in 1980. The USS *Enterprise* was powered by eight nuclear reactors fueled with HEU and was in service for over fifty years (1961–2012).
Source: U.S. Navy.

weapon-grade (93.5 percent uranium-235) uranium and, after 1962, with "super-grade" uranium (97.4 percent enriched)[42] made especially for the navy. With the end of U.S. production of HEU in 1992, it was decided to fuel naval reactors with weapon-grade uranium salvaged from surplus U.S. Cold War nuclear warheads.

Russia fuels its submarines with HEU but with a variety of enrichments. The cores of its second-generation reactors, many still in service in 2013, are enriched to 21 percent, and the cores of its third-generation reactors reportedly have zoned enrichment ranging from 21 to 45 percent. These reactors are designed to be refueled every eight years.[43] India has built its first nuclear-powered submarine with a fuel enrichment that is believed to be similar to that in Russia's third-generation reactors.[44]

France used HEU fuel in its first-generation ballistic-missile submarines but, after ending its production of HEU for weapons in the 1990s, decided that it would be more economical to fuel its naval reactors with LEU in the enrichment range it was already using in power reactors, that is, uranium enriched to 6 percent uranium-235 or less.[45] China is believed to use LEU as well.[46] Finally, Brazil, the first non-weapon state to launch a serious program to build a nuclear-powered submarine, has opted for 4 to 6 percent enriched LEU fuel—at least initially.[47]

The United States, Germany, and Japan all built prototype civilian nuclear-powered ships. As of 2013, only Russia operates nuclear-powered icebreakers and transport ships in the Arctic.[48] Three of Russia's nuclear-powered ships use a 135-MWt reactor fueled with 90 percent enriched uranium.[49] LEU fuel (slightly less than 20 percent) has been designed, however, for a variant of this reactor adapted for use in a barge-mounted nuclear power plant.[50] In 2011, Russia was reported to have decided to use this LEU fuel for a follow-on generation of nuclear icebreakers, the first of which is scheduled to begin operations in 2020.[51]

Converting Naval Reactors to Low-Enriched Fuel

In 1994, the U.S. Congress requested that the Office of Naval Nuclear Propulsion report on the possible "use of low-enriched uranium (instead of highly-enriched uranium) as fuel for naval nuclear reactors."[52] When the requested report came back in June 1995, it stated, in summary, that "the use of LEU in U.S. naval reactor plants is technically feasible, but uneconomic and impractical."[53] The report asserted that, unlike the case of research reactor fuel, "no more uranium could be packed into a modern long-lived [naval reactor] core without degrading the structural integrity or cooling of the fuel elements." If core life was to be preserved, therefore, it would be necessary to increase the volume of the fuel about three times.[54] It was estimated that, after one-time costs due to design change, the resulting annual cost increase—much of it apparently due to the cost of fabricating a greater volume of fuel—would be about $1 billion/year (1995 $).[55] The navy report emphasized that the volume of spent fuel to be managed also would increase threefold. It concluded that, "the use of LEU for cores in U.S. nuclear powered warships offers no technical advantage to the navy, provides no significant non-proliferation advantage, and is detrimental from environmental and cost perspectives."[56]

France arrived at a different conclusion. While the United States is moving to "lifetime cores" for its naval reactors,[57] France has switched to low-enriched fuel and refuels every seven to ten years.[58] The alternative of keeping naval reactors the same size but refueling more frequently was considered in the U.S. Navy report to Congress but the report concluded that the refuelings would keep the submarines and aircraft carriers in port for an extra two to four years over their lifetimes. In the case of the French submarines, refueling apparently does not add to overhaul times. This is attributed, at least in part, to the installation of large refueling hatches in the hulls of the French submarines.[59] U.S. submarines without lifetime cores are refueled by cutting a large hole into the hull and then welding it shut again after refueling.

The continued use of HEU for naval fuel by the United States, Russia, the United Kingdom, and India unnecessarily leaves open opportunities for theft of HEU. The associated huge Russian and U.S. stockpiles of excess weapon-grade uranium reserved for naval fuel also could become a barrier to continued progress in the reduction of stockpiles of weapon-usable fissile materials.

A small step toward ending the naval use of HEU came in early 2014. In response to a second request from Congress, the U.S. Department of Energy's Office of Naval Reactors for the first time indicated that a shift to LEU fuel might be practical, stating that "the potential exists to develop an advanced fuel system . . . [that] might enable either a higher energy naval core using HEU fuel or allow using LEU fuel with less impact on reactor lifetime, size and ship costs."[60] The new report suggested that developing the new fuel would be a long-term effort, with the implication that any construction of U.S. nuclear-powered submarines or ships with LEU cores may be decades in the future. All of the governments with HEU-fueled reactors, whether used for naval propulsion, military or civilian research, or for isotope production, should launch serious analyses of different strategies and the costs and benefits of ending their reliance on HEU fuel. These assessments should go beyond narrow naval requirements and include the benefits to international security from shifting to LEU fuel and shutting down unneeded HEU-fueled research reactors.

III

Eliminating Fissile Materials

8

Ending Production of Fissile Materials for Weapons

In December 1993, the United Nations General Assembly agreed, without a dissenting vote, on a resolution developed by a group of nineteen countries, including the United States and India and led by Canada, calling for "a non-discriminatory, multilateral and internationally and effectively verifiable treaty banning the production of fissile material for nuclear weapons or other nuclear explosive devices."[1]

It was expected that negotiations would begin at the United Nations Conference on Disarmament (CD) in Geneva after talks on the Comprehensive Test Ban Treaty (CTBT) were completed in 1996. As of the end of 2013, negotiations had not started on a Fissile Materials Cutoff Treaty. The hoped-for agreement is sometimes also referred to as a Fissile Material Treaty (FMT) or a Fissile Materials (Cutoff) Treaty to reflect different views on what it might cover. To inform discussions and support the start of talks at the CD, the IPFM has circulated a draft treaty text.[2]

Sixty-five countries participate in the CD and key decisions, such as establishing an agenda and a program of work for the year, must be agreed by consensus. For more than a decade, there was no consensus that negotiations on an FMCT should be given priority over negotiations on nuclear disarmament, a commitment to nonuse of nuclear weapons against non-weapon states, or prevention of an arms race in outer space. In 2009, there was for a brief period a consensus to launch discussions on those three issues while negotiating an FMCT.[3] Pakistan decided to block proceeding with this agreement, however, because it did not want to be locked into what it perceived was an inferior position relative to India with regard to its stocks of fissile materials.[4]

Unlike Pakistan, Israel is not publicly blocking the start of FMCT negotiations at the CD. Israel has made it clear, however, that it will not join such a treaty. Israel's prime minister Benjamin Netanyahu wrote President Clinton in 1999: "We will never sign the treaty, and do not delude yourselves—no pressure will help. We will not sign the treaty because we will not commit suicide."[5]

Israel is unique among the nuclear weapon states in having a policy of not acknowledging publicly that it has nuclear weapons. The verification arrangements for an FMCT would make that posture difficult to sustain.

For the United States, Russia, France, and the United Kingdom—all of which have declared that they have permanently ended their production of fissile materials for weapons—the rationale for an FMCT is to limit the nuclear weapon buildups of China, India, Pakistan, Israel, and North Korea. From the perspective of the non-weapon states, along with making the current fissile material production moratoria in four nuclear weapon states legally binding, a cutoff treaty would reduce the discriminatory nature of the NPT by requiring some monitoring of civilian nuclear operations in the weapon states. Finally, it could make irreversible the post–Cold War reductions in Russian and U.S. weapons stocks of fissile materials. Since the IAEA has long experience in inspecting civilian fissile material production facilities and stockpiles to provide confidence that they are not being diverted to weapons purposes, it is generally assumed that the IAEA would be the tasked with the verification of an FMCT.

The History of the FMCT

A cutoff was first proposed in 1957 on the personal initiative of President Eisenhower as a way to cap the Cold War nuclear arms buildup. The United States proposed "that an agreement be reached under which at an early date under effective international inspection, all future production of fissionable materials shall be used or stockpiled exclusively for non-weapons purposes under international supervision."[6] The Soviet Union, then far behind the United States in the production of plutonium and HEU, rejected the proposal and the nuclear arms race continued unabated.

With the end of the Cold War and the collapse of the Soviet Union in 1991, however, the situation changed radically. The United States, Russia, France, and the United Kingdom all decided to downsize their Cold War nuclear arsenals and as a result found themselves with far more fissile material than they needed. By 1996, each of the four governments had publicly declared that it had ended its production of HEU and plutonium for weapons. China did not make such a public declaration and did not

downsize its nuclear arsenal, but is believed to have suspended its production of fissile material for weapons in 1990.[7] China has turned an
underground site with three uncompleted plutonium production reactors
abandoned in the 1980s into a tourist attraction (figure 8.1).

Issues of the Scope of a Possible Treaty

The minimal undertakings of a cutoff treaty would be to not produce
fissile material for weapons and to decommission military fissile material
production facilities or convert them to non-weapon purposes. The
parties also would undertake not to separate plutonium or enrich
uranium except under safeguards and would be required to place under
safeguards any newly produced HEU or separated plutonium. A treaty
could possibly allow production of HEU, for example for naval propulsion reactor fuel, but only with arrangements to assure that it would not
be used for weapons.

HEU, plutonium, and uranium-233 are classified in the IAEA Statute
as "special fissionable materials," or materials that could be used to
make a fission explosive. The minor fissile materials neptunium-237
and americium-241, which have fast critical masses comparable to
uranium-235 (see figure 2.2), are considered "alternative nuclear materials" by the IAEA. They could be used to make a nuclear explosive
and, if separated, should therefore also be covered by safeguards under
an FMCT.[8]

Tritium, the heavy isotope of hydrogen with a 12.3-year half-life that
is used to boost the yield of modern nuclear warheads (see chapter 2),
is not a fissile material and would therefore not be captured under an
FMCT. A ban on tritium production, however, would make some FMCT
verification tasks easier since tritium is produced in reactor facilities.

An FMCT—or a treaty complemented by side agreements or unilateral commitments—could go beyond a ban on new production of fissile
materials for weapons to include also a ban on the use in weapons of
preexisting stocks of fissile material declared to be excess to all military
purposes, including material recovered from nuclear weapons and potentially including even HEU stocks reserved for naval reactor fuel.

It also would be natural to put under safeguards civilian nuclear material in the nuclear weapon states produced before the FMCT came into
force. In France and the United Kingdom, civilian fissile material already
is under Euratom safeguards and, in the United States, it has been offered
for IAEA safeguards. If preexisting civilian stocks were not put under

a

b

Figure 8.1

China's unfinished underground production reactor complex (2010). (a): Entrance to the complex at Baitao, east of Chongqing. The sign in Chinese above the tunnel reads "816 Underground Nuclear Project." (b): Project 816 reactor control room. Three of the four displays would have shown the core arrangements for three plutonium-production reactors that were never completed.

Source: Renren (top) and Sinaweibo (bottom).

safeguards, a complex dual accounting system would have to be established with new civilian material put under safeguards while preexisting civilian material was not.

The five NPT weapon states have already committed to place fissile material declared excess for military purposes under safeguards. At the 2000 NPT Review Conference, they pledged "to place, as soon as practicable, fissile material designated by each of them as no longer required for military purposes under IAEA or other relevant international verification and . . . to ensure that such material remains permanently outside of military programmes."[9] This commitment was not renewed at the 2010 NPT Review Conference, however.[10] One problem appears to be that, once material is placed under IAEA safeguards, those safeguards follow that material in perpetuity until the material is discarded as waste. Russia may have concerns about this approach because it plans to commercialize a breeder reactor fuel cycle in which plutonium would be recycled indefinitely. As a result, as of late 2013, there was still no agreement on IAEA monitoring of the 34 tons of plutonium that Russia and the United States each had declared excess under an agreement that they made in 2000 (see chapter 9).[11]

Russia did accept U.S. monitoring of the blend-down of 500 tons of its excess weapons HEU for sale to the United States, however, and the United States invited the IAEA to verify the blend-down of some of its own excess HEU.

France and the United Kingdom, which have significantly downsized their relatively more modest weapon stockpiles and already are comfortable with Euratom safeguards on their civilian fissile material, also could declare more of their weapons materials excess and place them under international monitoring.

India, however, apparently wants to keep its options open with regard to possible future weapons use of its civilian fissile material. During its negotiations with the United States in 2006 over exempting India from the Nuclear Suppliers Group ban on nuclear trade with non-parties to the Non-Proliferation Treaty, India refused to put its breeder reactor program under IAEA safeguards, arguing that plutonium in that program might be required for a "minimum credible deterrent."[12]

As disarmament proceeds, the large stocks of HEU reserved for naval fuel in the United States and Russia also could become of concern. As noted earlier, the United States has publicly reserved 152 tons of weapon-grade uranium for future use as naval reactor fuel.[13] This would be enough for at least 6,000 nuclear warheads. Russia may not yet have established a separate reserve of HEU for naval reactor use, but certainly will take that future need into account as it decides on further reductions of its military stock of HEU.

Some countries would like to see a treaty go still further and require reductions in existing stocks of weapons material. In 2010, Pakistan's ambassador to the CD insisted, "We could not accept a treaty that would freeze existing asymmetries or imbalance in fissile materials stockpiles between Pakistan and its neighbor which obviously had a head start."[14] In fact, although it started producing HEU a decade earlier than India, Pakistan only began to separate plutonium after about 1998 and has been making large investments in building up its plutonium production capacity. In 2013, it had two operating production reactors, another nearing operation, and a fourth under construction.[15] An early cutoff treaty would negate the value of these investments.

FMCT Verification Challenges

The verification arrangements for an FMCT would build on the monitoring techniques developed by the IAEA to verify that non-weapon states are complying with their NPT commitment not to produce "nuclear weapons or other nuclear explosive devices."[16] Adaptations to these techniques would be required, however, at enrichment and reprocessing plants that were not designed with safeguards in mind and enrichment plants that had previously produced HEU. It would also require "managed access" arrangements at military nuclear facilities in nuclear weapon states to allow verification while protecting sensitive information.

Under an FMCT, it would be necessary to verify

• non-operation of shutdown enrichment and reprocessing facilities;

• non-diversion of fissile materials produced at operating enrichment and reprocessing plants;

• the absence of clandestine production;

• non-production at suspect military nuclear facilities; and

• non-diversion of HEU from naval fuel cycles.

Non-operation of Shutdown Enrichment and Reprocessing Plants
Since the NPT weapons states ended or suspended production of fissile materials many years ago, most of their military fissile material production facilities have been shut down (table 8.1). Verifying that a facility is

Table 8.1
Shutdown enrichment and reprocessing plants

Enrichment plants

Country	Facility	Operating dates
China	Lanzhou	1964–1998
	Heping	1975–1987
France	Pierrelatte	1967–1996
	Georges Besse	1979–2012
United Kingdom	Capenhurst	1954–1982
United States	Oak Ridge, TN, electromagnetic	1945–1947
	Oak Ridge, TN	1945–1985
	Portsmouth, OH	1956–2001
	Paducah, KY	1952–2013

Reprocessing plants

Country	Facility	Operating dates
China	Jiuquan	1970–1984
	Guangyuan	1976–1990
France	Marcoule, UP1	1958–1997
Russia	Mayak, Ozersk BB Plant	1959–1987
	Seversk (Tomsk-7)	1956–2012
	Zheleznogorsk (Krasnoyarsk-26)	1964–2012
United Kingdom	Sellafield, B204	1952–1973
United States	Hanford, T and B Plants	1944–1956
	Hanford, REDOX Plant	1952–1967
	Hanford, PUREX Plant	1956–1990
	Savannah River, F Canyon	1954–2002
	West Valley, civilian	1966–1972

Sources: China: *Global Fissile Material Report 2010: Balancing the Books, Production and Stocks* (Princeton, NJ: International Panel on Fissile Materials, 2010), http://fissilematerials.org/library/gfmr10.pdf. France: ibid.; Mycle Schneider and Yves Marignac, *Spent Nuclear Fuel Reprocessing in France* (International Panel on Fissile Materials, April 2008). Russia: Thomas B. Cochran, Robert S. Norris, and Oleg Bukharin, *Making the Russian Bomb: From Stalin to Yeltsin* (Boulder, CO: Westview Press, 1995), 79–80. United Kingdom: *Global Fissile Material Report 2010*; Martin Forwood, *The Legacy of Reprocessing in the United Kingdom* (International Panel on Fissile Materials, July 2008). United States: R. E. Gephart, *A Short History of Hanford Waste Generation, Storage, and Release*, PNNL-13605 Rev. 4 (Richland, WA: Pacific Northwest National Laboratory, October 2003); "SRS History Highlights," www.srs.gov.

Notes: Enrichment plants are Gaseous Diffusion Plants except for the Y-12 electromagnetic plant at Oak Ridge that operated during World War II. While most of the plants were used to produce HEU for weapons, some produced only LEU. The Paducah plant produced LEU feed material for further enrichment to HEU at the Portsmouth plant. Reprocessing plants typically use the PUREX process to separate plutonium from spent nuclear fuel. The REDOX processes were used at the Hanford T, B, and REDOX Plants.

not operating can be done relatively non-intrusively, which would help assuage the concerns of countries such as China that do not yet wish to reveal how much HEU and plutonium they produced at these facilities.

Operating Enrichment and Reprocessing Plants
After an FMCT comes into force for a country, the IAEA will monitor its operating enrichment plants to determine whether or not they were producing HEU. There should be no requirements to produce new HEU for civilian purposes since HEU is being phased out for research reactors. There is ample excess Russian and U.S. HEU to fuel those research reactors still fueled with HEU during the transition. Russia and the United States also could fuel their naval propulsion reactors and the United States could continue to provide the United Kingdom with excess weapon-grade HEU for its nuclear submarines for several decades. India may be the only country with HEU-fueled submarines that does not have access to a large stockpile of HEU.[17]

One complication in verifying the non-production of HEU in enrichment plants that were used to produce HEU before an FMCT came into force would be that swipe samples at the facility are likely to find some particles containing HEU (figure 8.2). These particles would have to be convincingly attributed to pre-FMCT operations.[18] Failing that, another verification approach not dependent on swipe samples would have to be used.

In the case of reprocessing, the storage and use of newly produced plutonium, uranium-233, neptunium-237, and americium-241 would be subject to IAEA monitoring.

International safeguards at reprocessing plants currently are very costly. As of 2012, the IAEA had only two reprocessing plants under safeguards—both in Japan—but they accounted for roughly 20 percent of its safeguards budget.[19] In 2012, there were nine civilian reprocessing plants operating in the weapon states (see appendix 2).

Safeguards measurements on the plutonium flowing through Japan's reprocessing plants typically have one percent uncertainties, which would be enough to make ten nuclear weapons a year if the Rokkasho Reprocessing Plant operated at its design capacity of extracting 8 tons of plutonium per year. The real-time verification situation with the weapon-state reprocessing plants would be even worse. These plants were not designed for safeguards, and the IAEA did not have an opportunity to verify their design or install measuring devices before key areas became inaccessible because of high radiation levels. It may be difficult to measure the

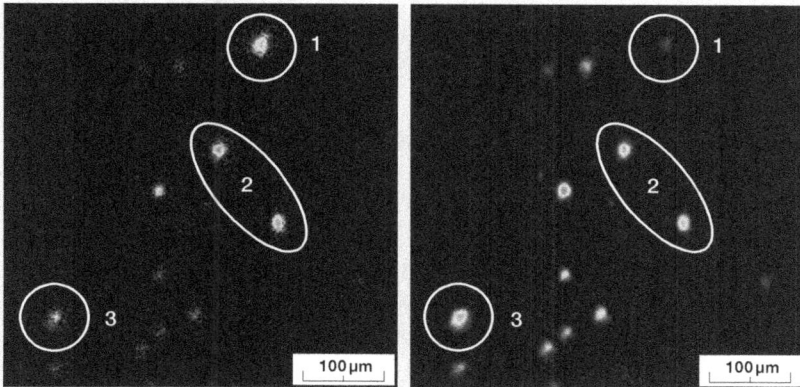

Figure 8.2

Images of micron-sized particles of uranium oxide taken with a Secondary Ion Mass Spectrometer. A beam of ions scans the particles, knocking out uranium ions whose masses are then measured in a mass spectrometer. The images at the left and right show respectively the uranium-235 and uranium-238 concentrations on the particle surfaces. The brightness of the particles in the images increases with the content of uranium-235 (left) and uranium-238 (right) but does not scale linearly with enrichment. Particles that are brighter in the left-hand image carry highly enriched uranium. Particle 1 is HEU, Particles 2 are of intermediate enrichment, and Particle 3 is probably natural uranium.
Source: Magnus Hedberg, European Commission Joint Research Centre: Institute for Transuranium Elements, Karlsruhe.

plutonium flows within these plants. Under an FMCT, they likely will have to be treated as "black boxes," with annual cleanouts and the IAEA comparing inputs with outputs.[20] Even better, of course, would be to abandon reprocessing, which, as discussed in chapter 6, is unnecessary and uneconomical.

Potential Clandestine Production

Although the weapon states would not join an FMCT if they foresaw the need to make more fissile material for weapons, the potential for clandestine production in the weapon states would exist in the longer term just as it does in the non-weapon states.

Over the past three decades, several non-weapon states party to the NPT have secretly begun construction of nuclear facilities that could have been used to make fissile material for weapon purposes. Iraq pursued an ambitious enrichment program after Israel destroyed its

Osirak reactor in 1981. More recently, in 2002 and 2009, Iran was discovered to be in the early stages of construction of the underground centrifuge enrichment plants at Natanz and Fordow, respectively. In 2007, Syria was found to have been building an undeclared reactor that seemed similar to North Korea's plutonium production reactor. All these activities were detected early on by other countries through intelligence gathering and led to demands for IAEA access and inspections.

In the non-weapon states, IAEA safeguards provide another level of protection by monitoring natural and low-enriched uranium that might be fed into an undetected enrichment plant, and spent fuel that might be fed into a clandestine reprocessing plant. The IAEA also has authority to visit suspect sites, which it has used in Iraq, North Korea, and Iran. First, however, it attempts to determine the nature of the facility from the outside and then, if necessary, by inspection of its interior.

One possible indicator of undeclared activities relating to gas centrifuge uranium enrichment would be unexplained evidence in the environment of degradation products of uranium hexafluoride (UF_6), the uranium-containing gas that is spun in the centrifuges. If it leaks out, UF_6 quickly reacts with water vapor in the air to become UO_2F_2, which is a solid and settles out.

Detection of clandestine reprocessing might be done with off-site environmental measurements to detect elevated levels of the volatile radioisotopes that are released when spent fuel is chopped and dissolved to extract the plutonium. Despite the advanced filtration systems in France's La Hague reprocessing plant, elevated levels of carbon-14 have been found in vegetation and elevated levels of iodine-129 in cattle thyroids in the surrounding area.[21]

When a reprocessing plant is operating, it also can be detected at a distance by observation of the plume of the fission product gas krypton-85 (half-life 11 years) that it releases. Figure 8.3 plots the reported weekly krypton-85 releases from Japan's Tokai reprocessing plant during the period 1995–2001 on the same timeline as measurements of the concentration of krypton-85 in the atmosphere at the Meteorological Research Institute in Tsukuba 55 kilometers away. This data suggests daily measurements of krypton-85 at Tsukuba would offer a greater than 90 percent probability of detecting the (protracted) separation at Tokai of 20 kilograms of weapon-grade plutonium or more per year.[22] This corresponds to 2.5 "significant quantities."[23] Production of smaller amounts of plutonium can be detected if the material is separated more quickly.

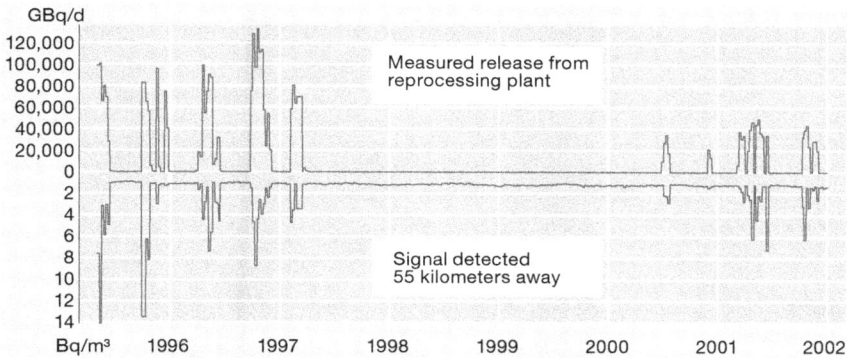

Figure 8.3

Remote detection of reprocessing. Reported releases of krypton-85 due to reprocessing at Japan's Tokai Pilot Reprocessing Plant (above the axis) on the same timeline scale as measurements of krypton-85 concentrations in the atmosphere of Tsukuba 55 kilometers away (below the axis).

The global background of krypton-85 from former military reprocessing and especially from ongoing civilian reprocessing may make finding signals from clandestine reprocessing more difficult (figure 8.4). Ending civilian reprocessing would in time sharply reduce the noise in the global krypton-85 background, making it easier to detect emissions from clandestine reprocessing facilities.

Managed Access in Military Nuclear Facilities

Preexisting fissile materials in nuclear weapon and naval nuclear fuel facilities would be exempt from IAEA monitoring under an FMCT. If there were grounds for suspicion that a facility was harboring clandestine enrichment or reprocessing activities, the IAEA would have to be able to mount a challenge inspection, whether or not the facility contained nuclear weapons, weapon components, naval reactor fuel, or related materials.

The weapon states would have sensitivities about letting foreign inspectors into such facilities but this situation is not unprecedented. Non-weapon states have sensitive military facilities, too.[24] Furthermore, the weapon states already have to be prepared for the possibility of challenge inspections at *any* facility by the Organization for the Prohibition of Chemical Weapons, the verification agency for the 1993 Chemical Weapons Convention (CWC). For sensitive sites, the CWC offers the

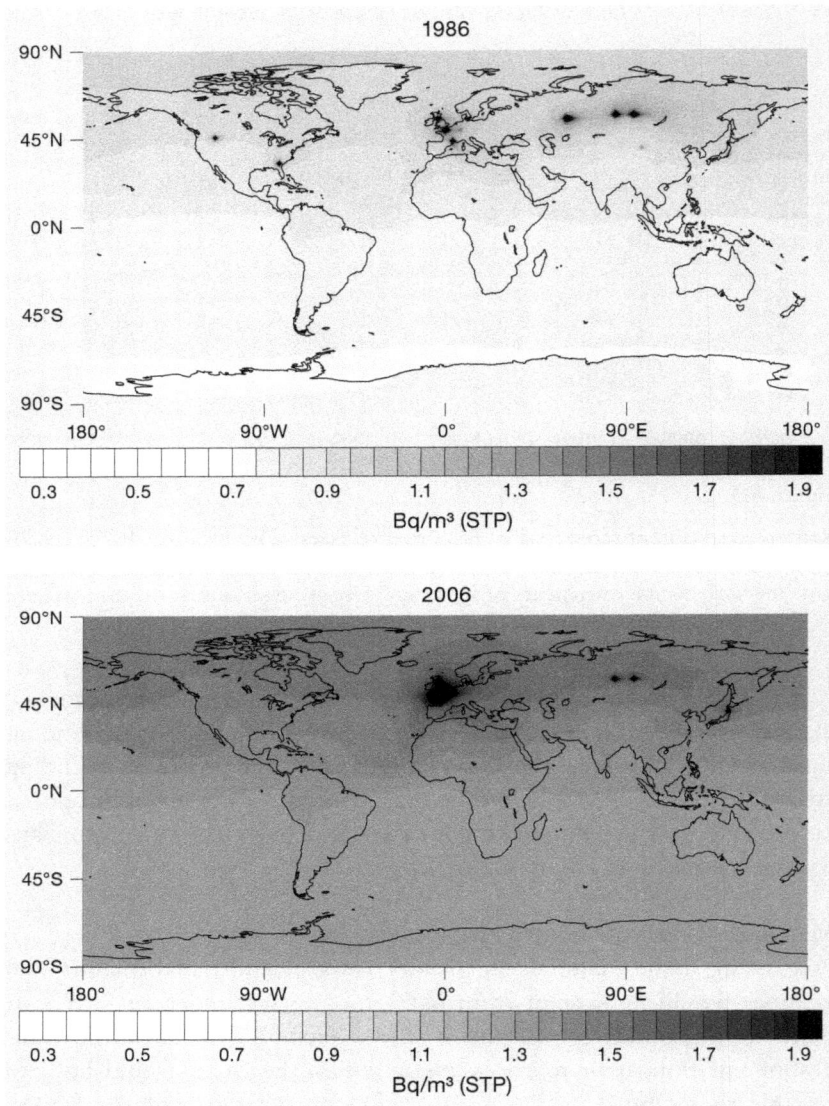

Figure 8.4

Calculated surface concentrations of krypton-85 in 1986 and 2006. These values are based on a global atmospheric-transport model, with source information based on actual and estimated emissions from several large reprocessing facilities that were operating at the time. The maps show the significant increase in global krypton-85 background due to large-scale reprocessing activities between 1986 (top) and 2006 (bottom). The U.S. government and possibly others have made extensive measurements (using airborne and ground-based air sampling) to estimate plutonium production in other countries.

Source: Ole Ross, "Simulation of Atmospheric Krypton-85 Transport to Assess the Detectability of Clandestine Nuclear Reprocessing," PhD thesis, University of Hamburg, 2010.

option of "managed access" under which the host government would be allowed measures to protect unrelated sensitive information.[25]

Under the CWC, inspectors are not allowed to analyze samples off site, and on-site analysis is allowed only for the presence of chemicals identified under the treaty as potential chemical weapon agents, their precursors, and degradation products. The inspectors can bring with them, however, an automated gas-chromatograph mass spectrometer whose database has been verified by the host country to contain only the indicators of those specified chemicals. The machine also has to operate in a "black-box" mode, namely, not reveal raw data from its measurements but only the conclusion from its internal computerized analysis of the data to determine whether any weapon-related chemicals have been detected.[26]

In the case of suspicions that clandestine centrifuge enrichment is taking place in a sensitive weapon-state military nuclear facility, inspectors might want to bring with them an instrument that could detect the degradation product UO_2F_2 from leaking uranium hexafluoride gas on the walls of the suspect facility. Uranium hexafluoride is not expected to be present in nuclear weapons production or naval fuel fabrication facilities.

An instrument designed to give a positive signal only if it detected uranium and fluorine at the same spot could detect the presence of UO_2F_2 without revealing the presence of other materials. Laser-induced breakdown spectroscopy, which has been adapted for IAEA use,[27] could do this. It involves using a laser to heat a small sample of material to temperatures at which it breaks down into atoms and excited ions. The atoms then release their internal energy in the form of light with wavelengths characteristic of each element that is present.

Naval Fuel Cycles

As discussed in chapter 7, at least four weapon states fuel their naval reactors with HEU. The United States and the United Kingdom use weapon-grade uranium (more than 90 percent enriched in uranium-235), and Russia and India use HEU enriched to a lesser degree.[28] If these countries continue to use HEU to fuel their naval reactors, it will become necessary eventually to produce more. But the use of HEU produced after the FMCT comes into force will have to be monitored to assure that it is not being diverted to weapons.

In principle, safeguards on the naval fuel cycle could be patterned on safeguards for the fuel cycles of civilian reactors, where the uranium is tracked from conversion to UF$_6$ through enrichment and fuel fabrication until it is installed and sealed in a reactor core—and later, after it is discharged from the core, through storage to reprocessing or disposal in a geological repository.

In the case of naval reactors, however, there are sensitivities about exposing reactor or fuel design information to foreign inspectors. Indeed, much less information has been declassified about U.S. naval reactor and fuel design than about U.S. nuclear weapon design.[29]

The IAEA has not yet had the challenge of verifying non-diversion from the naval fuel cycle because only the nuclear weapon states have nuclear-powered vessels and they are exempt from full-scope NPT safeguards. The situation is changing because Brazil, a non-weapon state, has decided to build nuclear submarines. Its reactors are to be LEU fueled—at least initially.[30] The IAEA monitors LEU use in non-weapon states to provide an added level of protection against the possibility that a country might have a small clandestine enrichment plant that could use the LEU as feed to make HEU.

The standard safeguards agreement for non-weapon states allows a country to remove nuclear materials from safeguards for "a non-proscribed military activity," notably naval propulsion.[31] This is a "loophole" in the NPT,[32] and it is quite possible that it could also be included in the Fissile Material Cutoff Treaty. The naval fuel cycle could then become a major problem for verifying nuclear disarmament.

A first step to reduce the size of the loophole would be to put stocks of HEU reserved for naval reactor use under IAEA safeguards. If much of the HEU is in excess weapons components as in the United States, it might first have to be converted to unclassified form.

A second step, if spent naval fuel is not reprocessed, as is the case in 2013 in all states with nuclear navies except for Russia, would be to put it under IAEA monitoring. In order to protect sensitive design information, the IAEA might have to measure the amount of HEU in the fuel by neutron interrogation from outside a container. If the spent naval fuel were reprocessed, under the FMCT, the recovered HEU would automatically come under IAEA safeguards.

Even with these measures in place, the amount of unsafeguarded HEU in naval fuel cycles would be huge. This would be especially true in the United States and the United Kingdom, where submarine naval reactors are designed with "life-time" cores, that is, the reactors are designed not to be refueled for thirty to forty years.[33] According to the U.S.

declaration of the history of its HEU stocks, the HEU inventory of the U.S. naval nuclear propulsion program as of 1996 was approximately 100 tons, and was mostly "in or has been used in naval cores [or] to be fabricated into fuel in the near future."[34]

Given the complexity and sensitivities of this situation, a possible way forward to devising acceptable and effective approaches to verifying that HEU is not diverted from naval fuel cycles would be through a cooperative research effort between safeguards and nuclear navy experts. These complications also provide yet one more reason for converting naval reactors to LEU fuel as discussed in chapter 7.

9

Disposal of Fissile Materials

Since the end of the Cold War, Russia and the United States have declared substantial quantities of their highly enriched uranium and plutonium excess to any military need and agreed to dispose of them. Much of the excess HEU already has been blended down to low-enriched uranium and preparations are being made to dispose of the excess plutonium. The United Kingdom, with a stockpile of approximately 100 tons of separated civilian plutonium—more than one-third of the global stockpile of separated civilian plutonium—also is actively studying how to dispose of its material.[1]

Much more Russian and U.S. weapons material could be declared excess and become subject to disposal. The United Kingdom and France could also declare excess substantial fractions of their more modest weapon-related stockpiles and arrange for their disposal. And, if Russia, the United Kingdom, and the United States designed new naval reactors to use low-enriched uranium—as France has done—Russian and U.S. naval reserves of HEU also could be phased out.

Factors Affecting Disposal Choices

In selecting disposal strategies, countries should take into account the degree of irreversibility being sought, materials security, cost, and international verifiability.

Degrees of Irreversibility

One fundamental difference between HEU and plutonium disposal is that HEU can be "denatured" isotopically while plutonium cannot. HEU can be blended with depleted uranium, natural uranium, or LEU to a level below 6 percent enrichment where the concentration of uranium-235 is too low to sustain an explosive chain reaction. To regain weapon-usable material from this low-enriched uranium requires re-enrichment of the uranium.

Isotopic denaturing is not practical for plutonium, since almost any mixture of plutonium isotopes can sustain an explosive chain reaction.[2] Currently, the dominant approach used for disposal of separated plutonium is to mix plutonium oxide with depleted uranium oxide to produce mixed oxide fuel. Irradiation of the MOX fuel in a reactor fissions a fraction of the plutonium and leaves the remainder mixed with highly radioactive fission products. Recovery of the residual plutonium in the spent MOX fuel would require remotely controlled chemical separation behind heavy radiation barriers. This level of disposal was described as "the spent fuel standard" in a 1994 U.S. National Academy of Sciences study on disposal of excess weapons plutonium because it would make the plutonium as inaccessible as that in spent LEU fuel.[3] The NAS study also noted that the spent fuel standard could be achieved by mixing the plutonium with some of the fission product wastes from which it had been originally separated and immobilizing the mixture in glass.

In principle, it would be possible to go beyond the spent fuel standard and devise means to fission separated plutonium almost completely.[4] This would be pointless, however, unless countries also were prepared to do this for the far greater quantities of plutonium in civilian power reactor spent fuel.

In a century or so, thirty-year half-life cesium-137, the fission product that generates most of the long-lived gamma radiation field around spent fuel, will have largely decayed away, and the spent fuel will no longer be self-protecting. At that point, irreversibility of disposal would depend upon the plutonium-bearing matrix being emplaced deep underground in a burial site under international surveillance.

Security

While making HEU or plutonium less accessible for weapons use is a critical long-term goal, processing the material instead of storing it could increase its exposure to potential theft in the short term. Countries therefore should not proceed to disposal until the associated arrangements are adequately secure. Conversely, disposal should not be delayed indefinitely. Even a small annual risk of insider theft, outsider penetration, or loss of government control could accumulate over a century into a significant risk.

The danger of terrorist acquisition of fissile material is particularly acute in the case of HEU. As noted in chapter 7, U.S. nuclear materials security experts believe that, if a terrorist group gained access to HEU in a storage facility, it might even be able to construct and detonate an "improvised nuclear explosive" on the spot before the guard force of the facility could stop them.

Plutonium would be harder for a terrorist group to fashion into a weapon, though prudently this should not be considered impossible. Also there is the possibility that a terrorist group could use stolen plutonium as a radiological weapon, dispersing it into the atmosphere with an explosive or with fire. If a large population inhaled a collective total of 0.1 grams of plutonium out of one kilogram dispersed in an urban atmosphere, on the order of one thousand extra cancer deaths could result.[5] The deaths would probably be invisible statistically but, as with the predicted 16,000 cancer deaths across Europe from the Chernobyl disaster, the psychological impact on the exposed population could be great.[6]

Cost

It is generally more economical to produce LEU by blending down existing HEU than by the normal process of enriching natural uranium (ignoring the sunk cost of making the HEU). Governments therefore can sell LEU blended down from excess HEU at a profit.

In contrast, because of the need to protect workers from the hazard of plutonium inhalation, it is more costly to fabricate plutonium into MOX fuel than to purchase the equivalent amount of low-enriched uranium fuel. Excess separated plutonium therefore is not an economically valuable energy resource but rather should be considered a dangerous waste to be disposed of. For nuclear energy establishments that have long believed that plutonium is the fuel of the future, however, treating plutonium as waste has been difficult to accept.

International Verifiability

The IAEA has the international responsibility to verify the acquisition and disposition of nuclear material in the non-weapon states. This monitoring responsibility continues for spent fuel even after it is disposed of in a deep underground geological repository. The IAEA does not have such responsibilities in the weapon states but is expected to acquire them as nuclear weapon reductions proceed. Ultimately, the weapon states may have to provide the IAEA a complete account of their fissile material production and disposition and facilitate the IAEA's verification of these declarations to the extent possible. This has already happened once with South Africa, which eliminated its small stockpile of nuclear weapons in

1990.[7] The task will be more difficult in weapon states where large amounts of fissile material have been disposed of in waste and consumed in nuclear weapon tests. These losses amounted to ton quantities both of HEU and plutonium in the cases of Russia and the United States.[8]

It may never be possible to reconstruct the histories of the Soviet/Russian and U.S. fissile material stockpiles precisely enough to exclude the possibility that one or both of them have hidden enough material to make up to hundreds of nuclear weapons, but those uncertainties could increase to the order of ten thousand warhead equivalents if there is no credible verification of the elimination of the stocks that Russia and the United States have declared excess.

The experiences of Russia and the United States with their strategies for disposing of their excess Cold War HEU and plutonium are described below.

Highly Enriched Uranium

In both Russia and the United States, excess HEU recovered from weapons is blended with natural or slightly enriched uranium to produce low-enriched uranium, which is then used in light water power reactor fuel.

Russia

In October 1991, just as the Soviet Union was disintegrating, Thomas L. Neff, an independent U.S. physicist and uranium market analyst, suggested that the United States incentivize the orderly disposal of Russia's excess HEU by offering to buy LEU derived from it.[9] Two years later, after a great deal of further effort by Neff and others, a contract was signed between Russia and the United States under which 500 tons of excess Russian HEU with an average enrichment of 90 percent would be blended down in Russia to LEU enriched to 4–5 percent and then be bought by the recently privatized U.S. Enrichment Corporation for resale to nuclear power utilities.[10] The rate of blend-down was limited to 30 tons per year so as not to disrupt world uranium and enrichment markets. Even at that rate, however, the flow of this LEU was enough to fuel about half of the U.S. nuclear power reactor fleet for fifteen years.[11]

Because of (since revised) U.S. industry concentration limits on the minor radioactive isotope uranium-234 in fuel, Russia diluted its weapon-grade uranium with 1.5 percent enriched uranium-235 blend stock produced from depleted uranium with a very low uranium-234 content.[12]

Production of this blend stock at the rate needed to blend down 30 tons of HEU annually required over five million separative work units per year—about the same amount that would have been required had the low-enriched uranium been produced directly from natural uranium. From the perspective of Russia's nuclear establishment, however, this arrangement still made sense because, in effect, it opened up the U.S. market for unused Russian uranium enrichment capacity.

In order to share the work from the HEU deal as widely as possible within Russia's nuclear complex, much of the weapon-grade uranium was shipped thousands of kilometers between the various stages of the process: reduction of the HEU metal to shavings, conversion into HEU oxide, conversion to HEU hexafluoride, and finally downblending.[13] The numerous long-distance shipments of HEU raised security concerns but the United States provided high-security railcars and no theft attempts have been reported.

As part of the blend-down agreement, Russia and the United States established bilateral verification measures. Inspectors from the United States made a total of twenty-four annual visits to the Russian facilities involved in down-blending and U.S. instrumentation at the HEU blend points monitored the blending process continuously. Russia also had the right to send inspectors to the United States to verify that the resulting LEU was fabricated into power reactor fuel and not reenriched.[14]

All of the 500 tons had been blended down by the end of 2013. Russia is believed to have hundreds of additional tons of excess HEU that could be disposed of in the same manner, but declined to extend the contract, preferring to keep the HEU as a domestic reserve.[15] One possible consideration is that much of Russia's remaining HEU contains reactor-produced isotopes of uranium because it was used as fuel in plutonium production reactors before being enriched. The LEU produced by down-blending this HEU therefore might not meet international standards.[16]

United States

The United States produced less HEU for weapons than the Soviet Union and has declared less excess. Furthermore, almost all of the weapon-grade uranium that the United States has declared excess for weapons purposes has been put into a reserve for future use in naval propulsion reactor fuel. About 181 tons of HEU, mostly less than weapon grade, was committed for blend-down. In addition, as of 2012 there were plans to reprocess two tons of spent research reactor fuel and blend-down the recovered HEU.[17]

The HEU-containing components of U.S. nuclear weapons are dismantled at the Department of Energy's Y12 complex in Oak Ridge, Tennessee. Most of the blend-down work has been contracted out to the Babcock and Wilcox Corporation, which also fabricates HEU fuel for the U.S. nuclear navy.[18]

As of the end of 2012, the United States had blended down 141 tons of its excess HEU. Between 1997 and 2012, the blend-down rate averaged about 10 tons per year. Looking forward, however, the rate of blend-down was expected to slow because of bottlenecks in the U.S. warhead and component dismantlement process. Completion of the blend-down of the remaining 42 tons was not committed to be completed until 2030.[19] The United States invited the IAEA to monitor some but not all of its HEU blend-down operations.

Separated Plutonium

In 2000, Russia and the United States concluded a Plutonium Management and Disposition Agreement (PMDA) that committed each to dispose of at least 34 tons of weapon-grade plutonium "withdrawn from nuclear weapon programs."[20] The United States also declared excess to its national security requirements an additional 20 tons of separated plutonium plus 7.8 tons of plutonium contained in government-owned spent fuel.[21] Included in the material declared excess for weapons purposes was all of the Department of Energy's non-weapon-grade plutonium (14.5 tons).[22]

The available disposal options for excess weapon-grade plutonium had been examined in the major study by the U.S. National Academy of Sciences referred to above. The study concluded that two disposal options would be the most feasible in the near term:

1. Immobilization of the plutonium in glass ("vitrification") along with fission product waste at one of the reprocessing facilities where the plutonium had originally been separated, and

2. Mixture of plutonium oxide into depleted uranium oxide, fabrication of the resulting mixed oxide into fuel, and irradiation of this fuel in U.S. light water power reactors.

The vitrified plutonium waste mixture or the spent MOX fuel then would be stored pending ultimate disposal in a geological repository.[23]

Russia's nuclear establishment objected to the United States adopting immobilization, however, since the plutonium would remain weapongrade. The United States therefore agreed that it would dispose of in MOX at least 25 of the 34 tons of plutonium subject to the disposition agreement.

The very high cancer risk from inhaled plutonium requires that MOX fabrication be carried out in glove boxes operated at less than atmospheric pressure (figure 9.1). The large quantity of plutonium involved also requires stringent material accounting, control, and physical security measures.

A gigawatt-scale LWR with a full core of MOX fuel could irradiate about 1 ton of weapon-grade plutonium annually.[24] Because their control systems are not designed for MOX fuel, however, most LWRs are limited to using only about one-third MOX fuel in their cores, which means that

Figure 9.1

A glove box at the UK Sellafield MOX fuel fabrication facility. The facility was closed in 2011 after it had been able to operate at an average of only one percent of design capacity for ten years. Plutonium handling operations take place in a shielded airtight glove box to protect workers from the hazard of plutonium inhalation.

Source: UK Nuclear Decommissioning Authority.

most 1 GWe reactors can irradiate annually only about one-third of a ton of weapon-grade plutonium.

Under the PMDA, disposal was supposed to begin by 2007 and quickly ramp up to a rate of 2 tons per year in each country. Because of delays, the agreement was amended in 2010. The date for beginning plutonium disposal was pushed back to 2018 and the minimum target rate of disposal was reduced to 1.3 tons/year.[25]

The disposal of Russian and U.S. plutonium was to be verified bilaterally and by the IAEA. According to both the original and amended PMDA, "Each Party shall have the right to conduct and the obligation to receive and facilitate monitoring and inspection activities . . . [and] shall begin consultations with the International Atomic Energy Agency (IAEA) at an early date and undertake all other necessary steps to conclude appropriate agreements with the IAEA to allow it to implement verification measures."[26] As of the end of 2013, however, a monitoring agreement had not yet been concluded.

Russia

The preference of Russia's nuclear establishment has always been to reserve all its civilian and excess weapons separated plutonium for its breeder program. Nevertheless, in 2000, it agreed to irradiate most of its excess weapons plutonium in MOX fuel for light water reactors if the United States and its allies supplied MOX fuel fabrication technology and fully funded the program.[27] Estimates of the construction costs of the U.S. and Russian MOX plants escalated in parallel, however,[28] while the funds that the United States and its allies were willing to provide for Russia's plutonium disposition program did not increase. As a result, in the 2010 amendments to the PMDA the United States agreed that Russia could irradiate its excess weapon-grade plutonium in its existing BN-600 prototype breeder reactor and in the BN-800 breeder reactor that was under construction.[29]

The United States pledged to provide $0.4 billion in support of this effort in exchange for a number of conditions. The central U.S. requirement was that Russia not separate plutonium from irradiated breeder fuel produced with the excess weapons plutonium until after all 34 tons of its excess weapons plutonium had been irradiated. This restriction on separating plutonium produced in the breeder extended to the plutonium generated in the uranium "blankets" around the breeder core, since neutrons emerging out of the core produce weapon-grade plutonium in these blankets.

United States

As has been noted, the United States initially adopted a two-track approach in which both the immobilization and MOX and fuel routes would be explored. The immobilization track became a "can-in-canister" approach in which the plutonium would first be embedded into ceramic cylinders inside metal cans. These cans would then be placed on racks inside large canisters, with the canisters being filled around them with molten glass mixed with fission product waste to provide a gamma radiation barrier. All this would be done at the Department of Energy's Savannah River Site in South Carolina where much of the U.S. weapons plutonium originally had been produced.

In 2002, since Russia insisted that the United States irradiate most of its plutonium, the U.S. Department of Energy decided that the two-track approach was too costly and canceled the immobilization track.[30] As of early 2013, however, the estimated cost of the U.S. MOX fuel fabrication facility had grown from $1 billion in 2002 to $7.7 billion. Projected costs for building and operating the facility and the associated Waste Solidification Building had ballooned from $2.2 billion to $18 billion, not including the cost of extracting the plutonium from excess weapons pits. The expected cost of operating the MOX plant had increased to $0.5 billion per year for fifteen years to dispose of the agreed 34 tons of plutonium.[31] The MOX fuel would replace about 700 tons of LEU fuel worth about $2 billion.

Which U.S. reactors would burn the MOX fuel remained uncertain. Because of utility concerns about the need for regulatory approval to use MOX instead of uranium fuel, questions outstanding about MOX fuel performance, and possible public opposition to its use, the Department of Energy offered utilities deep discounts for using the MOX fuel. Even so, after ten years of effort, the department was unable to find a private utility willing to use the MOX. Finally, in early 2013, the Obama administration announced that it was considering alternatives to plutonium disposal through use as MOX.[32]

The leading alternatives are

• dilution and disposal with plutonium-contaminated waste that is being disposed of in the Waste Isolation Pilot Plant in rooms carved out of a salt bed 650 meters under southeast New Mexico;

• a return to the can-in-canister approach of immobilization in high-level waste glass; and

• immobilization in ceramic and disposal in a deep borehole (see below).

Civilian Plutonium

The global stock of about 260 tons of separated civilian plutonium is mostly located at reprocessing plants in France, India, Japan, Russia, and the United Kingdom (figure 9.2). Its separation was originally launched in the expectation that it would be used as startup fuel for plutonium breeder reactors. In Russia and India, separated civilian plutonium is still reserved for this purpose. In France, however, in the absence of breeders, the separated plutonium is being recycled into MOX fuel for LWRs. Japan has long had similar plans and the United Kingdom has been focusing on that option as well.

In 2001, the United Kingdom completed a MOX plant at its Sellafield site to produce fuel for foreign reprocessing customers from the plutonium separated from their spent fuel. The plant was designed to produce 120 tons of MOX fuel per year but managed to produce only 14 tons of acceptable MOX fuel over the following decade.[33] In August 2011, the UK government therefore announced that the plant would be decommissioned.[34]

Shortly thereafter, however, the UK government announced a tentative decision to build a new MOX plant to dispose of its own plutonium.[35] It seems that despite the failed UK MOX fuel fabrication plant, the UK Department of Energy and Climate Change continued to favor the MOX option because of the successful operation of AREVA's MOX fuel fabrication plant in France.[36] There was, however, the as yet unresolved issue of MOX utilization in the United Kingdom. Currently, the United Kingdom has only one light water power reactor, which could irradiate only about half a ton of plutonium a year.[37] It is not yet certain that enough LWR capacity will be built there to irradiate 100 tons of plutonium in a reasonable length of time.

In parallel, however, the UK National Nuclear Laboratory was setting up a production line to immobilize chemically contaminated plutonium that otherwise would have required costly cleanup to make it pure enough for use in MOX fuel. The planned immobilization form was a mix of two ceramics, zirconolite and pyrochlore, which are resistant to long-term damage from radiation and to the leaching effects of water and so can hold the plutonium in place. The waste form was to be created by hot (1100–1320 °C) isostatic pressing at high pressure (1000 atmospheres) for eight to nine hours.[38]

By the time the United Kingdom is ready to make a final decision on plutonium disposal, the National Nuclear Laboratory project may have established immobilization as a credible option.

Figure 9.2
One of two plutonium storage facilities at the U.K.'s Sellafield reprocessing site.
The separate lockers prevent criticality accidents.
Source: IAEA ImageBank.

Direct Disposal Options
Relatively simple considerations suggest that direct disposal could be much less costly and complex than the MOX route. Direct disposal therefore deserves greater consideration.

MOX fuel pellets contain only about a gram of plutonium each and have to be machined precisely to fit snugly into the long zirconium tubes within which they are stacked. A huge number of such pellets would have to be produced—roughly one hundred million to dispose of 100 tons of plutonium. By contrast, a waste form designed for direct disposal could contain about one kilogram of plutonium and would not have to be machined to such precision.[39] The waste form also could be designed to be much more durable underground than spent MOX fuel. The direct disposal package could be buried in casks with spent fuel.

Another burial option would be to embed the plutonium in a low-leachable ceramic and dispose of it in boreholes at depths much greater than that of a geological repository for spent fuel (figure 9.3). Modern oil drilling techniques allow boreholes to go down to depths of up to five kilometers, ten times deeper than planned for a repository. This option has been studied for the disposal of both plutonium and spent fuel.[40] Deep borehole disposal of immobilized plutonium should be easier than spent fuel because the waste form could be more compact and robust and worker radiation doses would not be a concern due to the absence of strong gamma emitters.[41]

Several tons of plutonium could be disposed in a single borehole.[42] The appealing features of this concept are its simplicity, the difficulty of recovering the plutonium, and the possibility of remotely monitoring the site, including by commercial satellite imagery for any deep drilling activity. Also, the verification task would be simplified by the fact that the borehole would be open for a much shorter time than a mined geological repository.

Some U.S. separated plutonium already has been disposed of directly because it was so impure or diluted in other wastes that disposal in MOX would have been too difficult. As of the end of September 2009, the United States had disposed of an estimated 4.8 tons of plutonium in the Waste Isolation Pilot Plant (WIPP) in New Mexico.[43] WIPP is an operating deep underground repository licensed specifically for disposal of waste plutonium from the U.S. weapons program. In 2012, the U.S. Department of Energy proposed to dilute an additional 6 tons of plutonium to less than 10 weight percent concentration and ship it to WIPP.[44] As of the end of 2013 there had been no international monitoring of U.S. disposal of plutonium waste in WIPP.

The need for disposal options that are as irreversible as possible, secure, economical, and verifiable is likely to grow in importance. Much more weapons plutonium and HEU could be declared excess. As reported in chapter 4, the United States still has available for weapons enough HEU and plutonium to make 10,000 warheads, twice as much as required for the approximately 5,000 operational warheads that it declared at the 2010 NPT Conference. Russia probably has about the same number of

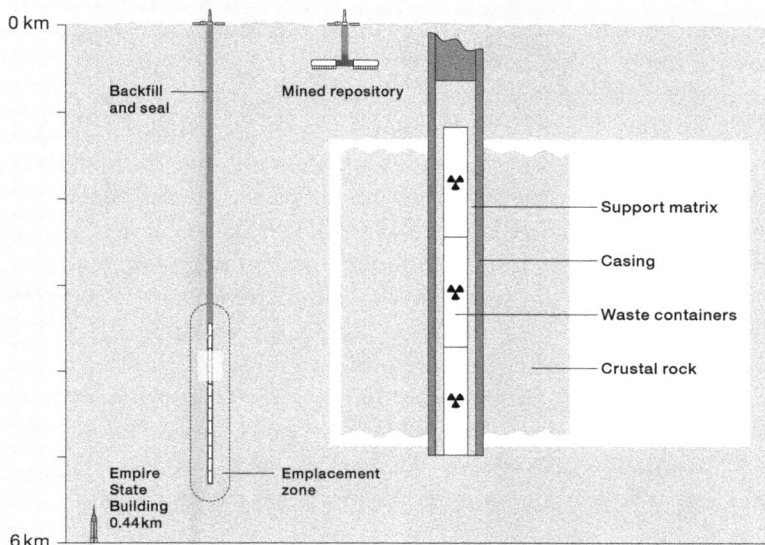

Figure 9.3

Schematic of deep borehole disposal of plutonium, high level radioactive waste, or spent fuel. A 5-kilometer deep borehole could be used for irreversible disposal of plutonium, high-level waste, or spent nuclear fuel with the material emplaced in containers at depths of over 3 kilometers. The space between the borehole casing and the containers would be filled with a dense material such as bentonite clay or cement. This filler material would help to keep water from reaching the emplaced waste. The upper 3 kilometers of the borehole would be sealed using clay, cement, and other material. The emplacement of plutonium and spent fuel could take place under international monitoring. Also shown is the typical 500-meter depth for a possible mined geological repository for nuclear waste and, for scale, the almost 400-meter tall Empire State Building in New York.

Source: Adapted from William G. Halsey, Leslie J. Jardine, and Carl E. Walter, *Disposition of Plutonium in Deep Boreholes*, Lawrence Livermore National Laboratory, UCRL-JC-120995, 1995; Fergus Gibb, N. A. McTaggart, K. P. Travis, D. Burley, and K. W. Hesketh, "High-Density Support Matrices: Key to the Deep Borehole Disposal of Spent Nuclear Fuel," *Journal of Nuclear Materials* 374 (2008): 370–377.

operational warheads but is estimated to have twice as much HEU and plutonium available for weapons as the United States.

Even with no further disarmament, the United States and Russia therefore could declare roughly half and three-quarters of their weapons materials excess respectively. Most of the remainder could be declared excess if they each reduced their warhead stockpiles from five thousand to one thousand or less. In early 2013 the United States indicated that it would seek agreement with Russia on further reductions in nuclear weapons, with a possible goal of 1,000–1,100 deployed strategic warheads.[45]

The challenge of disposing of civilian plutonium is also growing. Before the March 2011 Fukushima accident, for example, Japan was planning to recycle its separated plutonium in MOX fuel. Local safety concerns had already delayed the program by a decade, however. After the accident, the future of Japan's MOX program was even more uncertain. This situation provides yet another reason to develop a direct disposal alternative for plutonium and to stop separating more.[46]

10

Conclusion: Unmaking the Bomb

The production of fissile materials and the invention of nuclear weapons ushered in a new era in human affairs, one that held the potential for global catastrophe. In June 1945, with the world's first uranium enrichment plants beginning to produce tens of kilograms of highly enriched uranium and the first production reactors creating kilogram quantities of plutonium, a group of Manhattan Project scientists led by James Franck warned the U.S. government that "the development of nuclear power is fraught with infinitely greater dangers than were all the inventions of the past."[1]

Based at the University of Chicago, the scientists' group included Leo Szilard, the first to conceive of the nuclear chain reaction, and Glenn Seaborg, one of the discoverers of plutonium. They emphasized that the United States could not expect to have an enduring monopoly on fissile material and nuclear weapons: "We cannot hope to avoid a nuclear armament race either by keeping secret from the competing nations the basic scientific facts of nuclear power or by cornering the raw materials required for such a race. . . . It might take other nations three or four years to overcome our present head start, and eight or ten years to draw even with us if we continue to do intensive work in this field."[2]

Recognizing that countries might seek initially to develop the new nuclear technologies and materials in the hope of peaceful application, the scientists warned of the danger of nuclear weapons proliferation from "conversion of a peacetime nucleonics industry to military production." To forestall the risks of proliferation and nuclear arms racing, they proposed a ban on nuclear weapons and international control of the means of production of fissile material.[3]

These insights and proposals were developed further a year later in an official U.S. report on international control of atomic energy known as the Acheson-Lilienthal Report but authored in large part by Robert

Oppenheimer. The report argued that "national rivalries in the development of atomic energy readily convertible to destructive purposes are the heart of the difficulty" and warned that national control of uranium mining, uranium enrichment, and plutonium separation facilities would make it impossible to constrain nuclear weapons proliferation. It therefore proposed that fissile materials and all means of their production be placed under international ownership and control.[4]

The Cold War was already beginning, however, and it proved impossible in that context to realize the idea of international control. The Soviet Union tested its first nuclear weapon in 1949 (figure 10.1).

The runaway nuclear arms race between the United States and Soviet Union saw each building up arsenals of more than 30,000 nuclear weapons.[5] It ended after four decades, with Soviet leader Mikhail Gorbachev and U.S president Ronald Reagan agreeing at the 1986 Reykjavik summit on the goal of a world without nuclear weapons. Since then, Russian and U.S. nuclear arsenals have been reduced from tens of thousands to thousands of warheads. But over the course of the Cold War and the decades since it ended, the number of countries with nuclear weapons has grown to nine and the spread of enrichment and reprocessing technology as part of nuclear power programs has created additional "virtual" nuclear weapon states. The nuclear Non-Proliferation Treaty agreed on in 1968 has slowed but not stemmed the tide.

As of the end of 2013, the legacy of the Cold War and nuclear power programs was a global stockpile of about 1,900 tons of fissile material comprising about 1,400 tons of highly enriched uranium and 500 tons of separated plutonium. Since a modern version of the Nagasaki bomb design can use less than 4 kilograms of plutonium or 12 kilograms of highly enriched uranium, this global fissile material stockpile was enough for more than 100,000 weapons.

About 97 percent of the world's fissile material was owned by the nuclear weapon states. Over 900 tons were still in nuclear weapons or the associated production complexes, mostly in the United States and Russia, and about 180 tons of highly enriched uranium was assigned to naval propulsion programs. There was, however, not much pressure to further reduce these still huge stocks. The increased concern about nuclear terrorism following the September 11, 2001, attacks resulted primarily in reducing the number of HEU-fueled research reactors.

The earlier chapters of this book have mapped out the scale and nature of the global stockpiles of fissile material, the dangers they pose,

Figure 10.1

Leo Szilard (1898–1964), here in September 1949 reading about the onset of the U.S.-Soviet nuclear arms race, which he anticipated and warned against.
Source: Argonne National Laboratory, courtesy AIP Emilio Segrè Visual Archives.

and practical policy steps that could be pursued to drastically reduce the stocks and the use of fissile materials for nuclear reactor fuel. This final chapter provides an overview and summary plus a few additional suggestions.

There are two basic elements in our proposed approach to the reduction of the dangers from fissile materials: first, military and civilian fissile material stockpiles need to be capped and drastically reduced; and second, the regulation of fissile materials in all states should be approached as if the world is preparing for complete nuclear disarmament, however distant that goal may be. Progress on both will require the nuclear weapon states to demonstrate through their policies and budgets that they intend to deliver on their obligation to disarm. This would strengthen the nonproliferation regime and the determination of the global community to push back against the acquisition of nuclear weapon options through civilian nuclear power programs.

New initiatives in four major areas of fissile material policy are called for:

1. Increased transparency of nuclear warhead and fissile material stockpiles to facilitate deep reductions in arsenals;

2. A stop in further military and civilian production of HEU and plutonium and a phaseout of their use as reactor fuel;

3. Elimination of fissile material stockpiles in as irreversible a way as possible; and

4. Implementation of effective international verification of all of these steps.

Increasing Nuclear Transparency

Nonnuclear weapon states party to the NPT declare all their fissile and other nuclear materials to the International Atomic Energy Agency and these declarations are subject to IAEA verification. Since 1997, the five NPT weapon states (China, France, Russia, the United Kingdom, and the United States) have declared their civilian plutonium stocks annually and publically to the IAEA, with France and the United Kingdom also declaring their civil HEU stocks. Declarations of military stocks are much more uneven, however. The United States has made detailed declarations of its fissile material stockpiles and their respective production histories, including facility specific annual production data. The United Kingdom has made much less detailed declarations, reporting only the total size of its HEU and plutonium stockpiles. The other seven weapon states have made no declarations of their military stocks.

All nuclear weapon states need to become more transparent.[6] As a first step, each nuclear weapon state should publish its total holdings of HEU and plutonium as of specific recent dates and commit to subsequent annual updates of these holdings. In the longer term, all the weapon states should publish declarations covering the histories of their fissile material and warhead inventories. This would strengthen confidence in their commitment to verifiable nuclear disarmament and thereby strengthen the nonproliferation regime.

For those states just embarking on nuclear transparency, these measures could be implemented step by step, with states choosing initially to release information for either HEU or plutonium, or for one particular production site, or historical production data for one particular time period. Deeper reductions in existing nuclear warhead and fissile material stockpiles, however, will require sharing detailed and as complete data as can be reconstructed on the history of HEU and plutonium production and on existing and former production facilities.

Ending Production of HEU and Plutonium and Their Use as Fuel

Ending the production of fissile materials both for weapons and non-weapon uses is another critical step toward a world free of nuclear weapons and fissile material. With the end of the Cold War, the United States, the United Kingdom, France, and Russia all publicly announced that they had permanently stopped producing fissile materials for weapons. China has not made such a public declaration but has suspended production for more than two decades. Military production continues in India, Israel, Pakistan, and perhaps North Korea. Negotiations on a Fissile Material Cutoff Treaty were supposed to begin at the Conference on Disarmament in Geneva in 1996, but the required consensus on a negotiating agenda was blocked—sometimes by a single country. As of the end of 2013, negotiations had not begun.

The minimum obligations of states under a cutoff treaty would be not to produce fissile material for weapons and to decommission existing military fissile material production facilities or convert them to non-weapon purposes. Many uranium enrichment and reprocessing plants built by the NPT nuclear weapon states for the production of fissile material for weapons have already been shut down or decommissioned. The only operating facilities are in Israel, India, Pakistan, and possibly North Korea.

For a verified treaty, the parties also would undertake to accept international monitoring of their enrichment and reprocessing facilities and of any HEU and separated plutonium that they produced. The technical challenges of verifying an FMCT are largely similar to those used to verify the NPT in non-weapon states. The new challenges would be verifying non-diversion of HEU fuel from naval fuel cycles and establishing managed access arrangements for international inspectors at nuclear weapon and naval nuclear fuel fabrication sites to verify that no undeclared enrichment or reprocessing activities are taking place there.

Because of the huge stockpile of HEU from the dismantlement of excess stocks of Cold War weapons, there will be little need for Russia, the United States or the United Kingdom to produce additional HEU for naval reactor fuel for several decades. India had one operating HEU-fueled submarine, which was still undergoing trials in 2013, and planned to commission several more over the following decade. It should therefore be possible to have an international moratorium on HEU production for fuel purposes while the few countries that use HEU as a naval fuel convert to LEU fuel. A similar process has been working successfully for converting HEU-fueled research reactors to LEU fuel. These steps would lay the basis for a global phaseout of all HEU fuel use.

There is also a huge global stockpile of excess weapons plutonium from the downsizing of the Cold War arsenals and a still larger stockpile of civilian plutonium due originally to large-scale reprocessing in a few countries preparing for the deployment of plutonium-fueled breeder reactors. Even with the most optimistic growth projections for nuclear power, however, low-cost uranium will remain available for a century or longer. This means that the costly separation of plutonium for use as reactor fuel will continue to be noncompetitive.

Russia, China, and India are still conducting research and development on breeder reactors. But Russia could for decades rely on already separated plutonium for fueling its breeder prototypes. China has not yet made a commitment to a large-scale breeder program and hopefully will not do so. In India, breeder reactor advocates continue to control policy for nuclear energy research and development. Breeder programs in France and Japan have failed, but despite this it has proven difficult to reverse their reprocessing policies. Most countries, however, could back a moratorium on reprocessing for at least several decades and collectively press the holdouts to join suit.

There is a growing recognition that reliance on nuclear power will have important implications for the stability and verifiability of a world with many fewer nuclear weapons and much smaller fissile material stockpiles, and certainly for a nuclear weapon and fissile material free world. In time, a decision might be made to forego nuclear power entirely. As the Franck Report and later the Acheson-Lilienthal Report highlighted, in a disarmed world the existence of a nuclear power infrastructure would dangerously shorten the time for states to acquire nuclear weapons. Given this risk, it might be necessary for the world to do without nuclear power altogether.

Eliminating Stockpiles

A large amount of highly enriched uranium has been eliminated from the global stockpile due to the 1993 landmark agreement between the United States and Russia to blend down 500 tons of Russian weapon-grade HEU to LEU for power reactor fuel. This agreement was fulfilled in 2013. The United States eliminated over 140 tons of mostly non-weapon-grade HEU during the same period.

If the United States and Russia were to reduce their nuclear arsenals to 1,000 warheads each and if their nuclear navies shifted to LEU fuel, a further 1,000 tons of HEU could be declared excess and blended down to LEU.

Disposal of excess weapons plutonium has proven difficult and contentious. A total of 75 tons of weapons plutonium has been declared excess and still awaits disposal. The U.S.-Russian Plutonium Management and Disposition Agreement was signed in 2000 but, as of 2013, no material had yet been disposed of. The problem is much larger, however. The fact that the global inventory of civilian plutonium was about 260 tons in 2013, almost twice the amount of plutonium in weapon stockpiles, attests to the significance of the dangers from civilian plutonium utilization programs.

So far, the primary method for plutonium disposal is using it in mixed uranium plutonium oxide fuel for reactors. But, for most countries, this option has proven to be expensive and difficult. In the beginning of 2013, the United States called for a new look at the potential for direct disposal of immobilized plutonium. Other states with stocks of separated plutonium also should consider immobilization options. Placing the plutonium in several kilometer deep boreholes may be a viable option.

A Comprehensive Verification System

As countries embark on the nuclear disarmament path and undertake deep cuts in their nuclear arsenals, it will be critical to account for all their fissile materials and to establish comprehensive means for verifying its production and disposal. This will be an enormous challenge. Work on developing cooperative approaches to such verification is urgently needed.

One key task will be to develop reliable tags that can be applied to all intact nuclear warheads and to weapon components containing fissile material. The tags should allow the warheads and components to be tracked cooperatively through dismantlement to the final disposal of their fissile materials under international safeguards.[10]

Confidence also will have to be built that every warhead and kilogram of fissile material has been declared. Part of the answer will be "nuclear archaeology," an approach to verification that involves forensic analysis of former production sites and related records and waste materials.

To facilitate future verification of declarations of fissile material production, weapon states should commit to catalog and preserve operating records and waste materials and not undertake decommissioning of former fissile material production sites or dispose of wastes containing significant amounts of fissile materials until international inspectors have been able to characterize the sites and take samples for future analysis. Unfortunately, it is already too late for some sites but most have not yet been torn down.

As a start, weapon states could offer former fissile material production facilities as test beds for developing cooperative nuclear archaeology approaches and technologies and invite partners with similar facilities to join "site-to-site exercises" in verification approaches and techniques.

Verifying historical warhead and fissile material declarations—especially the huge Russian and U.S. stocks—will be an enormous undertaking. Because of the inevitable uncertainties in verifying past production and in ruling out the existence of undeclared warheads and secret facilities, technical verification will have to be complemented by other means to ascertain that all fissile material and nuclear weapons have been fully accounted for and that no more are being secretly produced. It may take many years for states to gain confidence in the correctness and completeness of each other's declarations.

One complementary approach to technical verification is called "societal verification." Leo Szilard first stressed its importance during World War II when he thought about possible international controls to prevent a post-war nuclear arms race between the United States and the Soviet Union.[12] One of his suggestions was to appeal to the conscience of scientists and engineers.[13] A global commitment to nuclear disarmament would create a context in which it would be reasonable to expect that scientists and engineers not work on secret nuclear weapon programs in their nations and blow the whistle on any such programs.[14] A vigorous proponent of societal verification in the context of verifying a nuclear weapon free world was the late Sir Joseph Rotblat (figure 10.2), a Manhattan Project scientist, one of the eleven who co-founded the scientists'

Figure 10.2

Joseph Rotblat (1908–2005), a Manhattan Project scientist, one of the founders of the scientists' Pugwash movement, and a strong advocate of societal verification. Rotblat, a Nobel Laureate, was a leading supporter of Israeli whistle-blower Mordechai Vanunu, arguing that Vanunu's exposure of Israel's nuclear weapon program was an act of conscience.

Source: Peter Hönnemann.

Pugwash movement to abolish nuclear weapons, and who urged in his 1995 Nobel Peace Prize speech that scientists "remember their responsibility to humanity."[15]

For societal verification to be effective, such reporting must be universally recognized to be the right and duty of all citizens. To this end, a nuclear disarmament treaty should contain a clause requiring that national laws be enacted that require and guarantee the right to report violations to international authorities.[16]

The change of perspective from national to global citizen envisaged by Szilard and Rotblat is not without historical precedent. The successful abolition of slavery, a legally established cornerstone of national economy and social values in many countries for hundreds of years, was achieved in considerable measure by a shift to the view that slavery was an affront to a common humanity and neither natural nor necessary. This success was won, however, only through sustained efforts over generations by anti-slavery activists and like-minded political leaders. The nuclear danger is immediate, however, and time is not on the side of humanity.[17]

Moving Beyond the Fissile Material Age

In a July 1944 letter to United States President Franklin D. Roosevelt, the Danish physicist Niels Bohr warned of the dangers of the coming fissile material age, urging that "unless . . . some agreement about the control of the use of the new active materials can be obtained in due time, any temporary advantage, however great, may be outweighed by a perpetual menace to human security."[18] Since then, a worldwide movement for nuclear disarmament has emerged. It has organized and educated fellow citizens about the nuclear danger and found allies in the leaders in many countries in the effort to create a nuclear weapons free world.[19]

There have been major successes. The very first resolution of the United Nations General Assembly, passed in January 1946, was a call for plans to abolish nuclear weapons. The pressure of international opinion has helped ensure that no nuclear weapon has been used in war since the destruction of Hiroshima and Nagasaki. The overwhelming majority of countries has chosen not to seek such weapons and more than half of the world's nations have joined nuclear-weapon-free zones that now cover Africa, Latin America, Central Asia, Southeast Asia, and the South Pacific. Signs of the continuing power of the impulse to seek nuclear disarmament were evident in the 2013 Oslo Conference on the

Humanitarian Impact of Nuclear Weapons, which brought together 128 states with the Red Cross Movement, United Nations agencies, and civil society to remind the world of the catastrophic effects of nuclear weapons use.

There is also growing international recognition of the acute dangers posed by fissile materials. The past two decades have seen governments mobilize to strengthen the nonproliferation regime by making nuclear programs more transparent and shutting down nuclear black markets. The international diplomacy surrounding Iran's nuclear program, backed by United Nations Security Council resolutions, is rooted in concern about the threat of latent proliferation stemming from the spread of uranium enrichment capabilities as part of the civilian nuclear fuel cycle.

The increased fear of nuclear terrorism has resulted in new initiatives to secure, consolidate and eliminate fissile materials seen to be vulnerable to theft. The heads of forty-seven national governments gathered at the first Nuclear Security Summit held in Washington in 2010 to focus on these issues. The second and third summits, in Seoul in 2012 and The Hague in 2014, drew fifty-three national leaders. A fourth summit was planned for 2016.

International efforts to abolish nuclear weapons and to prevent proliferation and nuclear terrorism have been acting largely in parallel. The fissile material perspective presented here provides them with a common basis. Banning nuclear weapons will not end the threat if countries continue to hold stocks of fissile material and use them in civilian programs. Ending the threat of proliferation and nuclear terrorism means recognizing that increasing security of fissile materials and controlling their production will not suffice indefinitely. Unmaking the bomb requires eliminating the fissile materials that make nuclear weapons possible.

Appendix 1: Enrichment Plants

Table A1.1
Major uranium enrichment facilities worldwide as of 2013

Facility	Type	Operational status	Safeguards status	Capacity (tSWU/yr)[1]
Brazil				
Resende	Civilian	Being commissioned	yes	115–200
China				
Shaanxi	Civilian	Operating	(yes)	1,000
Lanzhou II	Civilian	Operating	offered	500
Lanzhou III	Civilian	Operating	no	1,000
France				
George Besse II	Civilian	Operating	yes	7,500–11,000
Germany				
Gronau	Civilian	Operating	yes	2,200–4,500
India				
Ratehalli	Military	Operating	no	(15–30)
Iran				
Natanz	Civilian	Under construction	yes	8–120
Fordow	Civilian	Under construction	yes	5–10
Japan				
Rokkasho[2]	Civilian	Resuming operation	yes	50–1500
Netherlands				
Almelo	Civilian	Operating	yes	5,000–6,000

Table A1.1
(continued)

Facility	Type	Operational status	Safeguards status	Capacity (tSWU/yr)[1]
North Korea				
Yongbyon	?	Operating	no	8
Pakistan				
Kahuta	Military	Operating	no	(15–45)
Gadwal	Military	Operating	no	
Russia				
Angarsk	Civilian	Operating	no	2,200–5,000
Novouralsk	Civilian	Operating	no	13,300
Zelenogorsk	Civilian	Operating	no	7,900
Seversk	Civilian	Operating	no	3,800
United Kingdom				
Capenhurst	Civilian	Operating	yes	5,000
United States				
Eunice, NM	Civilian	Operating	offered	5,900

Sources: Enrichment capacity data is based on International Atomic Energy Agency (IAEA), Integrated Nuclear Fuel Cycle Information Systems (INFCIS), http://www-infcis.iaea.org; *Global Fissile Material Report 2013: Increasing Transparency of Nuclear Warhead and Fissile Material Stocks as a Step toward Disarmament* (Princeton: International Panel on Fissile Materials, 2013).

Notes: All plants listed are centrifuge uranium enrichment plants. Most of the HEU in the global stockpile was produced in gaseous diffusion plants, as of the end of 2013 these had all been shut down. Laser uranium enrichment technology is still under development.

[1]Capacity is given in ton separative work units (tSWU) per year, a measure of the capacity of plants to enrich uranium. Individual machines in these plants may have capacities ranging from 1–2 to 50–100 kg-SWU per year. Where the plant capacity is shown as a range, the lower value is current capacity as of 2013 and the higher value is planned capacity. Capacity shown in brackets is an estimate.

[2] In 2013, the Rokkasho centrifuge plant was being refitted with new centrifuge technology and was operating at very low capacity.

Appendix 2: Reprocessing Plants

Table A2.1
Major reprocessing facilities worldwide as of 2013

Facility	Type	Operational status	Safeguards status	Capacity (tHM/yr)[1]
China				
Lanzhou pilot plant	Civilian	Starting up	(no)	50–100
France				
La Hague UP2	Civilian	Operating	yes (Euratom)	1,000
La Hague UP3	Civilian	Operating	yes (Euratom)	1,000
India[2]				
Trombay*	Military	Operating	no	50
Tarapur*	Dual-use	Operating	no	100
Kalpakkam*	Dual-use	Operating	no	100
Israel				
Dimona*	Military	Operating	no	40–100
Japan				
Rokkasho	Civilian	Starting up	yes	800
Tokai	Civilian	Temporarily shut down	yes	200
North Korea				
Yongbyon	Military	On standby	no	100–150
Pakistan				
Nilore*	Military	Operating	no	20–40
Chashma*	Military	Under construction	no	50–100

Table A2.1
(continued)

Facility	Type	Operational status	Safeguards status	Capacity (tHM/yr)[1]
Russia				
Mayak RT-1	Civilian	Operating	no	200–400
United Kingdom				
B205[3]	Civilian	To be shut down	yes (Euratom)	1,500
THORP[4]	Civilian	To be shut down	yes (Euratom)	1,200
United States				
Savannah River H-Canyon	Civilian	Special operations	no	15

Sources: Data on design capacity is based on International Atomic Energy Agency (IAEA), Integrated Nuclear Fuel Cycle Information Systems (INFCIS), http://www-infcis.iaea.org; and IPFM, *Global Fissile Material Report 2013: Increasing Transparency of Nuclear-warhead and Fissile-material Stocks as a Step toward Disarmament* (Princeton, NJ: International Panel on Fissile Materials, 2013).

Note: *Processes heavy water reactor fuel. All others process LWR fuel except for B205, which processes fuel from graphite-moderated, carbon dioxide–cooled Magnox reactors.

[1]Design capacity refers to the highest amount of spent fuel the plant is designed to process and is measured in tons of heavy metal per year (tHM/yr), tHM being a measure of the amount of heavy metal—uranium in these cases—that is in the spent fuel. Actual throughput is often a small fraction of the design capacity. For example, Russia's RT-1 plant has never reprocessed more than 130 tHM/yr and France, because of the non-renewal of its foreign contracts, reprocesses 1050 tHM/yr. LWR spent fuel contains about 1 percent plutonium, and heavy-water and graphite-moderated reactor fuel about 0.4 percent.

[2]As part of the 2005 Indian-U.S. Civil Nuclear Cooperation Initiative, India has decided that none of its reprocessing plants will be opened for IAEA safeguards inspections.

[3]The last Magnox reactor was scheduled for shut down in 2014. The UK Nuclear Decommissioning Authority projected that the B205 plant would be shut down in 2016 after it completed reprocessing the remaining Magnox fuel.

[4]In July 2012 the British Nuclear Decommissioning Authority announced the planned closure of its Thorp reprocessing plant at Sellafield, after it completed its existing reprocessing contracts, expected in 2018.

Notes

1 Introduction

1. President Truman was informed about the Hiroshima bombing while on board the USS *Augusta* on his way back from the Potsdam Conference, which had ended on August 2, 1945. Richard Rhodes, *The Making of the Atomic Bomb* (New York: Simon & Schuster, 1995), 734.

2. "White House Press Release on Hiroshima, Statement by the President of the United States," August 6, 1945.

3. Ibid.

4. "Establishment of a Commission to Deal with the Problems Raised by the Discovery of Atomic Energy," Resolution 1.1, General Assembly United Nations, January 24, 1946.

5. Otto Frisch and Rudolph Peierls, "On the Construction of a Super-bomb Based on a Nuclear Chain Reaction in Uranium, Memorandum" (Birmingham University, March 1940). Reprinted in Robert C. Williams and Philip L. Cantelon, eds., *The American Atom: A Documentary History of Nuclear Policies from the Discovery of Fission to the Present, 1939–1984* (Philadelphia: University of Pennsylvania Press, 1984).

6. Quoted in Rhodes, *The Making of the Atomic Bomb*, 496.

7. J. Carson Mark, Theodore Taylor, Eugene Eyster, William Maraman, and Jacob Wechsler, "Can Terrorists Build Nuclear Weapons?," in Paul Leventhal and Jonah Wechsler, eds., *Preventing Nuclear Terrorism* (Lexington, MA: Lexington Books, 1987).

8. Remarks by President Obama at Hankuk University, Seoul, Republic of Korea, March 26, 2012.

9. The Atoms for Peace initiative was launched with a speech by President Eisenhower to the General Assembly of the United Nations on December 8, 1953.

10. Luis W. Alvarez, *Adventures of a Physicist* (New York: Basic Books, 1987), 125.

11. Oleg Bukharin and William Potter, "Potatoes Were Guarded Better," *Bulletin of the Atomic Scientists*, June 1995.

12. Matthew Wald and William J. Broad, "Security Questions Are Raised by Break-in at a Nuclear Site," *New York Times*, August 7, 2012; William J. Broad, "The Nun Who Broke Into the Nuclear Sanctum," *New York Times*, August 10, 2012.

2 Production, Uses, and Stocks of Fissile Materials

1. Frederick Soddy, "Radium," in A. T. Moore, ed., *Professional Papers of the Corps of Royal Engineers, Volume XXIX, 1903* (Chatham: Royal Engineers Institute, 1904), 251–252.

2. H. G. Wells, *The World Set Free* (London: Macmillan, 1914).

3. For a biography of Szilard, see William Lanouette, *Genius in the Shadows: A Biography of Leo Szilard: The Man Behind the Bomb* (New York: C. Scribner's Sons, 1992).

4. Reprinted in ibid., 205–206.

5. Niels Bohr and John A. Wheeler, "The Mechanism of Nuclear Fission," *Physical Review* 56 (September 1, 1939).

6. A third isotope (uranium-234) is a decay product of uranium-238 and is present in an extremely low concentration (0.0055 percent) in natural uranium.

7. Otto Frisch and Rudolph Peierls, "Memorandum on the Properties of a Radioactive 'Super-bomb'" (Birmingham University, March 1940). The authors recognize that the idea of uranium enrichment may not have occurred to the Germans yet so that their report should be kept secret for the time being.

8. The fission cross section of uranium-235 for fast neutrons was estimated to be 10^{-23} cm^2; it is however about ten times smaller, that is, closer to 10^{-24} cm^2.

9. Heisenberg's various estimates and the debate about them are discussed in Thomas Powers, *Heisenberg's War: The Secret History of the German Bomb* (New York: Knopf, 1993), 447–451.

10. This is the established usage of the term "fissile" in the nuclear arms control literature. In contrast, in nuclear engineering, an isotope is considered "fissile" if it fissions (and sustains a nuclear chain reaction) for neutrons of any energy, namely, for both fast and thermal neutrons. Isotopes that undergo fission only for high-energy neutrons are considered "fissionable." In other words, uranium-235 is fissionable and fissile; uranium-238 is only fissionable.

11. David L. Clark and David E. Hobart, "Reflections on the Legacy of a Legend: Glenn T. Seaborg, 1912–1999," *Los Alamos Science*, 2000, 56–61.

12. Uranium-233 is produced by neutron capture by thorium in a nuclear reactor, while both neptunium-237 and americium-241 are produced, like plutonium, by irradiating uranium fuel in a reactor. While uranium-233 weapons have been tested, there are no indications that uranium-233 is used in operational nuclear weapons today.

13. Stockpile estimates for separated neptunium are very difficult. Neptunium-237 constitutes about 6 percent of the plutonium inventory in spent fuel,

or about 0.07 percent of total heavy metal. In principle, separation of neptunium is straightforward, but the material is typically discarded with the waste stream of a reprocessing plant. As of March 31, 1998, the U.S. Department of Energy owned 466 kilograms of neptunium-237, of which 351 kilograms were in separated form. "Facsimile to David Albright from the Office of Declassification Security Affairs, U.S. Department of Energy, April 14, 1998," cited in David Albright and Kimberly Kramer, *Neptunium 237 and Americium: World Inventories and Proliferation Concerns (Revised)* (Institute for Science and International Security, August 22, 2005).

14. Edward Teller, *The Legacy of Hiroshima* (Garden City, NY: Doubleday, 1962).

15. For uranium conversion processes, see Manson Benedict, Thomas H. Pigford, and Hans Wolfgang Levi, *Nuclear Chemical Engineering*, 2nd ed. (New York: McGraw-Hill Book Company, 1981), 236–274.

16. For the scientific basis and history of isotopic separation using thermal diffusion, see G. Müller and G. Vasaru, "The Clusius-Dickel Thermal Diffusion Column—50 Years After Its Invention," *Isotopenpraxis Isotopes in Environmental and Health Studies* 24, nos. 11–12 (1988): 455–464.

17. Vincent C. Jones, *Manhattan: The Army and the Atomic Bomb* (Washington, DC: Center of Military History, U.S. Army, 1985), 172–183.

18. By September 1945, one set of calutrons at the Oak Ridge site had produced more than 88 kilograms of material with an average enrichment of 84.5 percent. Ibid., 148.

19. David Holloway, *Stalin and the Bomb* (New Haven: Yale University Press, 1994).

20. Mahdi Obeidi and Kurt Pitzer, *The Bomb in My Garden: The Secret of Saddam's Nuclear Mastermind* (Hoboken, NJ: Wiley, 2004). The discovery in the early 1990s of Iraq's calutron program by inspectors of the International Atomic Energy Agency, with the assistance of veterans of the U.S. Manhattan Project electromagnetic enrichment program, is described in Leslie Thorne, "IAEA Nuclear Inspections in Iraq," *IAEA Bulletin*, no. 1 (1992): 16–24. The consolidated IAEA reports of the inspections in Iraq were submitted to the United Nations Security Council as "Letter dated 6 October 1997 from the Director General of the International Atomic Energy Agency to the Secretary-General," United Nations Security Council S/1997/779, October 8, 1997.

21. In early 1940, Simon had the idea of using a thin porous metal barrier as the separation membrane. Nancy Arms, *A Prophet in Two Countries: The Life of F. E. Simon* (Oxford: Pergamon Press, 1966), 109.

22. Holloway, *Stalin and the Bomb*. See also Pavel V. Oleynikov, "German Scientists in the Soviet Atomic Project," *Nonproliferation Review* 7, no. 2 (Summer 2000): 1–30.

23. For a history see R. Scott Kemp, "The End of Manhattan: How the Gas Centrifuge Changed the Quest for Nuclear Weapons," *Technology & Culture* 53, no. 3 (July 2012). The quote is from F. A. Lindemann and F. W. Aston, "The

Possibility of Separating Isotopes," *Philosophical Magazine* 37, no. 221 (1919): 523–534.

24. Richard G. Hewlett and Oscar E. Anderson, *The New World: A History of the United States Atomic Energy Commission, Volume 1, 1939–1946*, California Studies in the History of Science (University of California Press, 1962), 49–107. The centrifuge project was formally canceled in January 1944. For a more detailed history, see B. C. Reed, "Centrifugation During the Manhattan Project," *Physics in Perspective* 11, no. 4 (2009): 426–441.

25. For a more detailed account, see R. Scott Kemp, "Nonproliferation Strategy in the Centrifuge Age" (PhD thesis, Princeton University, 2010).

26. Allan S. Krass et al., *Uranium Enrichment and Nuclear Weapon Proliferation* (London and New York: Taylor & Francis Ltd., 1983), 127 and 134.

27. Reconfiguration of the cascades would be preferable for efficient HEU production but would involve additional delays before production of weapon-grade uranium could begin.

28. Alexander Glaser, "Characteristics of the Gas Centrifuge for Uranium Enrichment and Their Relevance for Nuclear Weapon Proliferation," *Science & Global Security* 16, nos. 1–2 (2008): 1–25.

29. Both cases assume typical tails depletion levels of 0.3 percent.

30. Rodney P. Carlisle and Joan M. Zenzen, *Supplying the Nuclear Arsenal: American Production-Reactors, 1942–1992* (Baltimore, MD: Johns Hopkins University Press, 1996), 26–45.

31. E. A. G. Larson, *A General Description of the NRX Reactor*, AECL-1377 (Chalk River, Ontario: Atomic Energy of Canada Limited, July 1961), www .fissilematerials.org/library/lar61.pdf; Donald G. Hurst, *Canada Enters the Nuclear Age: a Technical History of Atomic Energy of Canada Limited* (Montreal: McGill-Queen's University Press, 1997).

32. Small amounts of plutonium-238 are also produced, mostly by multiple neutron captures, starting with a non-fission capture on uranium-235.

33. The Hiroshima bomb contained 64.1 kilograms of highly enriched uranium according to a Researchers Brief quoting A. Francis Birch, *Report of Gun Assembled Nuclear Bomb*, October 6, 1945. Birch's report itself is apparently no longer in circulation. For more details, see John Coster-Mullen, *Atom Bombs: The Top Secret Inside Story of Little Boy and Fat Man* (Waukesha, WI: John Coster-Mullen, 2005), 121.

34. John Malik, *The Yields of the Hiroshima and Nagasaki Nuclear Explosions*, LA-8819 (Los Alamos National Laboratory, 1985).

35. The Hiroshima bomb was not tested before it was used. See also the discussion of South Africa's nuclear program in chapter 3.

36. *Manual for Protection and Control of Safeguards and Security Interests*, DOE-M-5632.1C-1 (Washington, DC: U.S. Department of Energy, Office of Security Affairs, Office of Safeguards and Security, July 15, 1994).

37. Some "tactical" (battlefield) weapons used the gun-type method. See, for example, Thomas B. Cochran, William M. Arkin, and Milton M. Hoenig, *Nuclear Weapons Databook, Volume I: U.S. Nuclear Forces and Capabilities* (Cambridge, MA: Ballinger Publishing Company, 1984).

38. "On the Goals and the Program of Tests at the Test Site No. 2 in 1953" (Council of Ministers of the USSR Decision of 1953, 1953). For an analysis, see Pavel Podvig, "Interesting Document about Soviet Nuclear Tests in 1953," *Russian Strategic Nuclear Forces*, 2012, russianforces.org/blog/2012/10/interesting_document_on_soviet.shtml.

39. *Restricted Data Declassification Decisions 1946 to the Present (RDD-8)* (U.S. Department of Energy, January 1, 2002), 70.

40. The United States first demonstrated the principle of boosting in its 15th nuclear test (Operation "Greenhouse," Shot "Item") in April 1951, *United States Nuclear Tests, July 1945 Through September 1992*, DOE/NV-209, Revision 15 (U.S. Department of Energy, Nevada Operations Office, December 2000).

41. J. Carson Mark, "Explosive Properties of Reactor-Grade Plutonium," *Science & Global Security* 4, no. 1 (1993).

42. *Nonproliferation and Arms Control Assessment of Weapons-Usable Fissile Material Storage and Excess Plutonium Disposition Alternatives*, DOE/NN-0007, 37–39 (Washington, DC: U.S. Department of Energy, January 1997), www.fissilematerials.org/library/doe97.pdf.

43. Ibid.

3   The History of Fissile Material Production for Weapons

1. David Albright, Frans Berkhout, and William Walker, *Plutonium and Highly Enriched Uranium 1996: World Inventories, Capabilities and Policies* (Oxford: Oxford University, 1997). More recently, annual updates by IPFM, in particular *Global Fissile Material Report 2010: Balancing the Books, Production and Stocks* (Princeton, NJ: International Panel on Fissile Materials, 2010), www.fissilematerials.org/library/gfmr10.pdf.

2. Along the way, it also became clear that weapons design for plutonium would be much more challenging than originally anticipated, and the development of the implosion-assembly mechanism absorbed most of the intellectual efforts that went into bomb design between 1943 and 1945. Lillian Hoddeson et al., *Critical Assembly: A Technical History of Los Alamos During the Oppenheimer Years, 1943–1945* (Cambridge University Press, 1993).

3. For detailed discussion of the U.S. enrichment complex, see *Highly Enriched Uranium: Striking a Balance. A Historical Report on the United States Highly Enriched Uranium Production, Acquisition, and Utilization Activities from 1945 through September 30, 1996* (Washington, DC: U.S. Department of Energy, 2006), www.fissilematerials.org/library/doe06f.pdf.

4. See *Plutonium: The First 50 Years. United States Plutonium Production, Acquisition and Utilization from 1944 through 1994*, DOE/DP-0137

(Washington, DC: U.S. Department of Energy, February 1996), www
.fissilematerials.org/library/doe96.pdf.

5. Ibid.

6. *The United States Plutonium Balance, 1944–2009* (Washington, DC: U.S.
Department of Energy, June 2012), www.fissilematerials.org/library/doe12.pdf.

7. *Highly Enriched Uranium: Striking a Balance. A Historical Report on the
United States Highly Enriched Uranium Production, Acquisition, and Utilization
Activities from 1945 through September 30, 1996.* This was updated as
*Highly Enriched Uranium Inventory: Amounts of Highly Enriched Uranium in
the United States* (Washington, DC: U.S. Department of Energy, 2006), www
.fissilematerials.org/library/doe06f.pdf.

8. The definitive accounts on the Soviet effort to acquire nuclear weapons are
David Holloway, *Stalin and the Bomb* (New Haven: Yale University Press, 1994);
and Michael D. Gordin, *Red Cloud at Dawn: Truman, Stalin, and the End of
the Atomic Monopoly* (New York: Farrar, Straus and Giroux, 2009). For a com-
pilation of official records of the Soviet nuclear weapons program, see *Atomnyi
Proekt SSSR: dokumenty i materialy [The USSR Atomic Project: Documents and
Materials]*, ed. L. D. Ryabev (Moscow: Nauka-Fizmatlit; Sarov: RFYaTs-VNIIÉF:
1999–2010).

9. Holloway, *Stalin and the Bomb*, chap. 7.

10. Henry DeWolf Smyth, *Atomic Energy for Military Purposes: The Official
Report on the Development of the Atomic Bomb under the Auspices of the
United States Government, 1940–1945* (Princeton: Princeton University Press,
1945).

11. For example, the Smyth Report specified the main design choices made for
the reactors—such as the use of cylindrical uranium slugs in a square graphite
lattice—and also revealed the effective plutonium production rate of about 1
gram of plutonium per day for a power level of 500–1,500 kilowatts thermal
(ibid., §6.41).

12. The balance between openness and secrecy chosen for the Smyth Report was
the subject of intense debate. Rebecca Press Schwartz, "The Making of the
History of the Atomic Bomb: Henry DeWolf Smyth and the Historiography of
the Manhattan Project" (PhD thesis, Princeton University, 2008). The publication
of the report served several purposes, including demarcating technical information that would be considered "secret" (everything not mentioned in the report)
once the war ended.

13. Holloway, *Stalin and the Bomb*, chap. 9.

14. Robert Chadwell Williams, *Klaus Fuchs, Atom Spy* (Cambridge, MA:
Harvard University Press, 1987).

15. The following discussion on Russia's enrichment complex is based on *Global
Fissile Material Report 2010*, chap. 4A, which relied on Russian literature that
had only recently been published.

16. *Global Fissile Material Report 2010*, chap. 4.

17. This is a summary of a much more detailed discussion in *Global Fissile Material Report 2010*, chap. 3.

18. Twelve reactors were designed to produce plutonium and two to produce tritium and other isotopes.

19. The estimated stock of 128 tons of weapon-grade plutonium includes the plutonium produced since 1994 at the ADE-2 reactor at Krasnoyarsk and its counterparts, ADE-4 and ADE-5 in Seversk. These three production reactors produced a total of about 15 tons of weapon-grade plutonium after 1994, but Russia agreed with the United States that the plutonium in spent fuel produced after that date would not be used for weapons.

20. *Global Fissile Material Report 2011: Nuclear Weapon and Fissile Material Stockpiles and Production* (Princeton, NJ: International Panel on Fissile Materials, 2012), www.fissilematerials.org/library/gfmr11.pdf, fig. 6.

21. *Communication Received from the Russian Federation Concerning Its Policies Regarding the Management of Plutonium*, IAEA, INFCIRC/549/Add.9/15, May 23, 2014.

22. The Frisch-Peierls Memorandum and many other key historic documents are reproduced in Robert C. Williams and Philip L. Cantelon, eds., *The American Atom: A Documentary History of Nuclear Policies from the Discovery of Fission to the Present, 1939–1984* (Philadelphia: University of Pennsylvania Press, 1984).

23. Ibid. The memorandum estimated the amount of uranium-235 needed to "about 1 kg as a suitable size for the bomb," which is about one-fiftieth the actual value required without implosion, due primarily to an overestimate of the uranium-235 fission cross-section used in the estimate of critical mass.

24. *Report by M.A.U.D. Committee on the Use of Uranium for a Bomb, July 1941*, reproduced in Margaret Gowing, *Britain and Atomic Energy 1939–1945* (London: Macmillan and Co., 1964), 394–426. The report included an estimate of the cost of a nuclear weapons program able to produce 36 bombs per year, the design and cost of a gaseous diffusion isotope separation plant, and an estimate of the expected yield of the bomb. There was a parallel study on nuclear power, *Report by M.A.U.D. Committee on the Use of Uranium as a Source of Power*, June 1941. It is reproduced in Gowing, *Britain and Atomic Energy 1939–1945*, 427–436.

25. Status issues were still evident in British policymaking when the acquisition of thermonuclear weapons was considered in the mid-1950s. The UK 1956 Statement on Defense argued that, "our status as the great power is directly connected with the possession of the H-bomb," *Statement on Defense 1956*, Cmd 9691 (London: HMSO, 1956).

26. Margaret Gowing and Lorna Arnold, *Independence and Deterrence: Britain and Atomic Energy, 1945–1952* (New York: St. Martin's Press, 1974).

27. For a detailed account of the accident, see Lorna Arnold, *Windscale, 1957: Anatomy of a Nuclear Accident* (New York: St. Martin's Press, 1992).

28. In 1981, the name of the site was changed from Windscale back to its original name Sellafield.

29. The United Kingdom's official account of plutonium production says, "the military and civil nuclear cycles have been run in parallel and to some extent were entwined during the early years of the nuclear programme," *Plutonium and Aldermaston: A Historical Account* (UK Ministry of Defense, 2000).

30. *Global Fissile Material Report 2010*, chap. 5. For the official declaration of the UK HEU stockpile, see *Historical Accounting for UK Defense Highly Enriched Uranium* (UK Ministry of Defense, March 2006), www.fissilematerials .org/library/mod06.pdf.

31. *Plutonium and Aldermaston: A Historical Account*. Ironically, this official declaration is no longer available on the government website (www.gov.uk); it can be found at www.fas.org and is also mirrored at www.fissilematerials.org/ library/mod00.pdf.

32. This surplus material included only 0.3 tons of weapon-grade plutonium. All numbers are from "Supporting Essay Five: Deterrence, Arms Control, and Proliferation," §26 in *The Strategic Defense Review*.

33. The 1998 *Strategic Defense Review* states: "All stocks of highly enriched uranium will, however, be retained outside safeguards, since material no longer needed for nuclear weapons will be used for the naval propulsion programme," §26 in ibid.

34. "Agreement between the Government of the United States of America and the Government of the United Kingdom of Great Britain and Northern Ireland for Cooperation on the Uses of Atomic Energy for Mutual Defense Purposes," 1958. For a detailed discussion, see Jennifer Mackby and Paul Cornish, *U.S.-UK Nuclear Cooperation after 50 Years* (Washington, DC: Center for Strategic and International Studies, 2008). A comprehensive collection of interviews conducted for this project is available online at www.csis.org/program/us-uk -nuclear-cooperation-after-50-years. A National Security Directive from 1991 signed by U.S. President George H. W. Bush mandates that the Department of Energy "shall produce additional weapons parts as necessary for transfer to the United Kingdom pursuant to the Agreement of Cooperation," "National Security Directive 61, FY 1991–1996 Nuclear Weapons Stockpile Plan" (The White House, Washington, DC, July 2, 1991).

35. Details about the transfers of HEU and plutonium between the United States and the United Kingdom were omitted from the U.S. fissile material reports of 1996/2012 and 2001/2006. In 1962, the United States successfully conducted a nuclear test using power reactor plutonium that it had received from the United Kingdom under the 1958 Agreement. *Additional Information Concerning Underground Nuclear Weapon Test of Reactor-Grade Plutonium* (Washington, DC: U.S. Department of Energy, Office of the Press Secretary, June 27, 1994).

36. Prior to 1980, the United Kingdom bartered 5.4 tons of separated plutonium for 7.5 tons of HEU and 6.7 kilograms of tritium from the United States. *Plutonium and Aldermaston: A Historical Account*. For a more complete discussion, see *Global Fissile Material Report 2010*, chap. 5.

37. William Walker, *Nuclear Entrapment: THORP and the Politics of Commitment* (London: Institute for Public Policy Research, 1999).

38. In 2009, the UK Government began public discussion of its plutonium-disposition options. This led to a 2011 report *Management of the UK's Plutonium Stocks: A Consultation Response on the Long-Term Management of UK-Owned Separated Civil Plutonium* (UK Department of Energy and Climate Change, December 1, 2011).

39. *Communication Received from the United Kingdom of Great Britain and Northern Ireland Concerning Its Policies Regarding the Management of Plutonium, Statements on the Management of Plutonium and of High Enriched Uranium*, INFCIRC/549/Add.8/16 (International Atomic Energy Agency, June 18, 2013).

40. "En 1958 Marcoule produira 100 kg de plutonium" (In 1958, Marcoule will produce 100 kg of plutonium), *Sciences et Avenir* 109 (March 1956): 128–132; and also "Le drame du plutonium" (The plutonium drama), *Sciences et Avenir,* 135 (May 1958): 231–236.

41. The Eisenhower Administration refused to back the attempted seizure of the Suez Canal by Britain, France, and Israel. Richard K. Betts, *Nuclear Blackmail and Nuclear Balance* (Washington, DC: Brookings Institution, 1987), 62–66. Bertrand Goldschmidt quoted an agreement signed in the aftermath of the crisis under which "the CEA was to carry out preparatory research into atomic explosions and, should the government then decide to proceed further, preliminary research leading to the production of prototypes and the staging of tests." Bertrand Goldschmidt, *The Atomic Complex: A Worldwide Political History of Nuclear Energy* (La Grange Park, IL: American Nuclear Society, 1982), 137.

42. "Arrêt de La Production de Matières Fissiles Pour Les Armes Nucléaires," 2010, www.francetnp2010.fr/spip.php?article7.

43. "Pierrelatte: L'usine D'enrichissement de L'uranium, En Démantèlement, Juin 2009," 2010, www.francetnp.fr/spip.php?article67.

44. *Global Fissile Material Report 2010*, chap. 6.

45. "Speech by President Nicolas Sarkozy, Presentation of 'Le Terrible' in Cherbourg," March 21, 2008, www.fissilematerials.org/library/sar08.pdf.

46. Mycle Schneider and Yves Marignac, *Spent Nuclear Fuel Reprocessing in France* (International Panel on Fissile Materials, April 2008).

47. UP stands for *usine de plutonium* or "plutonium factory."

48. France operated six gas-cooled graphite power reactors and exported another one to Spain (Vandellos-1). It is generally assumed that some of the plutonium produced by these reactors also entered the French military stockpile.

49. *Communication Received from France Concerning Its Policies Regarding the Management of Plutonium, Statements on the Management of Plutonium and of High Enriched Uranium*, IAEA, INFCIRC/549/Add.5/17, August 28, 2013.

50. John Lewis and Xue Litai, *China Builds the Bomb* (Stanford, CA: Stanford University Press, 1988), 62.

51. *The Chinese Communist Atomic Energy* Program, U.S. National Intelligence Estimate Number 13–2-60, December 13, 1960, www.foia.cia.gov/docs/DOC_0001095912/DOC_0001095912.pdf.

52. Lewis and Litai, *China Builds the Bomb*, 65.

53. *Modern China's Nuclear Industry*, selections translated in *JPRS Report: Science & Technology, China*, JPRS-CST-88-002 (Springfield, VA, January 15, 1988), www.fissilematerials.org/library/jprs88.pdf.

54. Ibid.

55. This is made up of three centrifuge plants in Shaanxi (0.2 million SWU/yr installed in 1996, 0.3 million SWU/yr added in 1998, and 0.5 million SWU/yr added in 2011) and one built in Lanzhou in 2001 with a capacity of 0.5 million SWU/yr.

56. "China's Indigenous Centrifuge Enrichment Plant," *Uranium Intelligence Weekly*, October 25, 2010; Phillip Chaffee and Kim Feng Wong, "China's Indigenous Capacity May Be Double Previous Estimates," *Nuclear Intelligence Weekly*, March 1, 2013.

57. *JPRS Report: Science & Technology, China.*

58. Ibid.

59. *Global Fissile Material Report 2010*, 104.

60. The uncertainty of ±25 percent stems primarily from the uncertainty of the power levels of the two reactors.

61. *Global Fissile Material Report 2010*, 106.

62. Personal communication to Frank von Hippel, April 11, 1991. See also Albright, Berkhout, and Walker, *Plutonium and Highly Enriched Uranium 1996: World Inventories, Capabilities and Policies*, 38, 68, 76, 80.

63. *Banning the Production of Fissile Materials for Nuclear Weapons: Country Perspectives on the Challenges to a Fissile Material (Cutoff) Treaty* (International Panel on Fissile Materials, September 2008), 7–13, www.fissilematerials.org/library/gfmr08cv.pdf.

64. "Successful Hot Test of China's Pilot Reprocessing Plant," *IPFM Blog*, December 21, 2010.

65. *Communication Received from China Concerning Its Policies Regarding the Management of Plutonium*, IAEA, INFCIRC/549/Add.7/12, September 26, 2013.

66. The series of Sino-Soviet accords, which were crafted at about the same time (1955–1957), would have been similar in scope had cooperation not collapsed in 1959.

67. Interview with Francis Perrin (High-Commissioner, *Commissariat à l'Energie Atomique*, CEA, 1951–1970), "France Admits It Gave Israel A-Bomb," *Sunday Times*, October 12, 1986.

68. Yossi Melman, "Hollywood Producer Gave Israel Sketches of Centrifuges for Dimona Nuclear Reactor," *Haaretz*, July 19, 2011. See also William J. Broad,

John Markoff, and David E. Sanger, "Israeli Test on Worm Called Crucial in Iran Nuclear Delay," *New York Times*, January 15, 2011.

69. In particular, is difficult to make an unambiguous estimate of the reactor's power level based on Vanunu's account; see *Global Fissile Material Report 2010*, chap. 8.

70. According to anonymous U.S. officials, the thermal power of the Dimona reactor was probably increased from about 40 MWt shortly after it went critical in December 1963 to about 70 MWt prior to 1977 when Vanunu began working at Dimona.

71. Alexander Glaser and Marvin Miller, "Estimating Plutonium Production at Israel's Dimona Reactor" (presented at the 52nd INMM Annual Meeting, Palm Desert, CA, 2011). See also *Global Fissile Material Report 2010*, chap. 8.

72. Seymour M. Hersh, *The Samson Option: Israel's Nuclear Arsenal and American Foreign Policy* (New York: Random House, 1991), 180–181.

73. Sasha Polakow-Suransky, *The Unspoken Alliance: Israel's Secret Relationship with Apartheid South Africa* (New York: Pantheon Books, 2010), 122–123.

74. About 463 kilograms of HEU was "material unaccounted for" at the U.S. naval nuclear fuel plant in Apollo, Pennsylvania, during 1957–1978, its period of HEU operations. Most of the inventory difference occurred during the period 1957–1968 when the Nuclear Materials and Equipment Corporation (NUMEC) operated the plant. About 126 kilograms were eventually discovered during decommissioning of the plant, leaving 337 kilograms still missing. Victor Gilinsky and Roger J. Mattson, "Revisiting the NUMEC Affair," *Bulletin of the Atomic Scientists* 66, no. 2 (April 2010). Grant F. Smith, *Divert!: NUMEC, Zalman Shapiro and the Diversion of US Weapons Grade Uranium Into the Israeli Nuclear Weapons Program* (Washington, DC: Institute for Research, Middle Eastern Policy, 2012).

75. R. Scott Kemp, "Nonproliferation Strategy in the Centrifuge Age" (PhD thesis, Princeton University, 2010), chap. 3.

76. The First Atoms for Peace Conference in Geneva in 1955, which was chaired by Homi Bhabha, the founder of India's nuclear program, included significant sharing of information on nuclear reactor design and reprocessing chemistry.

77. Robert Bothwell, *Nucleus: The History of Atomic Energy of Canada Limited* (Toronto: University of Toronto Press, 1988), 350–371.

78. Canada apparently received assurances from India that the CIRUS reactor and its spent fuel would only be used for peaceful purposes, George Perkovich, *India's Nuclear Bomb: The Impact on Global Proliferation* (Berkeley: University of California Press, 1999), 27. On India's interpretation of its sovereign rights over the fissile material produced in reactors, see Itty Abraham, *The Making of the Indian Atomic Bomb: Science, Secrecy and the Postcolonial State* (New York: Zed Books, 1998).

79. In addition to the two Canadian PHWRs, India has four other foreign built reactors: two U.S.-supplied Boiling Water Reactors that began operating in 1969

and two 1,000 MWe Soviet/Russian light water reactors, the first of which began operating in 2013 and the second of which was expected to become operational in 2014. All these reactors are under IAEA safeguards. India also committed, as part of the U.S.-India deal, to place eight of its 220 MWe reactors under safeguards.

80. See *Global Fissile Material Report 2010*, chap. 9.

81. An estimated 240 kilograms of plutonium have been separated from the spent fuel of safeguarded PHWRs and are not available for weapons. They are not included in this estimate.

82. In order to achieve equilibrium-core operation for the PFBR as quickly as possible, the operator could have an incentive to fuel the reactor with reactor-grade plutonium rather than recycle the weapon-grade plutonium produced in the blanket. Alexander Glaser and M. V. Ramana, "Weapon-Grade Plutonium Production Potential in the Indian Prototype Fast Breeder Reactor," *Science & Global Security* 15, no. 2 (2007): 85–105.

83. PREFRE-1 at Tarapur (100 ton capacity, commissioned in 1977) and KARP at Kalpakkam (100-ton capacity, commissioned in 1998) treat spent fuel from the heavy water power reactors.

84. PREFRE-2 at Tarapur and Kalpakkam I and II each have a capacity to reprocess 100 tons of spent fuel per year. *Annual Report 2010–11* (Mumbai: Government of India, Department of Atomic Energy, 2011).

85. Saurav Jha, "Enrichment Capacity Enough to Fuel Nuke Subs," *IBNLive*, November 2011. Meena Menon, "Reprocessing of Spent Fuel Key to Nuclear Power Programme: Manmohan," *The Hindu*, January 7, 2011.

86. Jha, "Enrichment Capacity Enough to Fuel Nuke Subs."

87. Ibid.

88. Shahid-Ur Rehman, *Long Road to Chagai* (Islamabad: Printwise Publications, 1999), 50.

89. A. Q. Khan worked for *Fysisch Dynamisch Onderzoek,* a subsidiary of *Verenigde Machine Fabrieken,* a company that worked closely with *Ultra-Centrifuge Nederland* (UCN), the Dutch member of the Urenco uranium enrichment consortium. See for example, David Albright, *Peddling Peril: How the Secret Nuclear Trade Arms America's Enemies* (New York: Free Press, 2010).

90. "N-Capability Acquired in 1983, Says Qadeer," *Dawn*, May 30, 1999. Rauf Siddiqi, "Khan Boasts Pakistan Mastered Uranium Enrichment by 1982," *Nucleonics Week*, May 20, 1999.

91. *Nuclear Black Markets: Pakistan, A. Q. Khan and the Rise of Proliferation Networks* (London: International Institute for Strategic Studies, May 2007). Albright, *Peddling Peril.*

92. Simon Henderson, "Nuclear Scandal: Dr. Abdul Qadeer Khan," *Sunday Times*, September 20, 2009; R. Jeffrey Smith and Joby Warrick, "A Nuclear Power's Act of Proliferation," *Washington Post*, November 13, 2009.

93. See, for example, Leonard S. Spector, *Nuclear Proliferation Today* (New York: Vintage Books, 1984). *Nuclear Black Markets*.

94. Unis Shaikh and M. A. Mubarak, "Radiation Safety around the 'Hot Facilities' at the PINSTECH," *The Nucleus* 8, no. 4 (1971): 13–27.

95. SGN was responsible for the process engineering work at the pilot reprocessing plant, while Belgonucléaire designed the building.

96. David Albright and Paul Brannan, *Pakistan Expanding Plutonium Separation Facility Near Rawalpindi* (Institute for Science and International Security, May 19, 2009).

97. As much as 95 percent of the design plans may have been transferred, along with equipment, by SGN before it finally ended its role in the project in June 1979. *Nuclear Black Markets*.

98. David Albright and Paul Brannan, *Chashma Nuclear Site in Pakistan with Possible Reprocessing Plant* (Institute for Science and International Security, January 2007).

99. "Pakistan's Indigenous Nuclear Reactor Starts Up," *The Nation*, April 13, 1998.

100. Mark Hibbs, "After 30 Years, PAEC Fulfills Munir Khan's Plutonium Ambition," *Nucleonics Week*, June 15, 2000.

101. Mark Hibbs, "Bhutto May Finish Plutonium Reactor without Agreement on Fissile Stocks," *Nucleonics Week*, October 6, 1994.

102. Milton Benjamin, "Pakistan Building Secret Nuclear Plant," *Washington Post*, September 23, 1980.

103. "Pakistan is Reprocessing Fuel Rods to Create Plutonium Nuclear Weapons," *CBS News Transcripts* (6:30 PM ET), March 16, 2000. For further details, see Zia Mian and A. H. Nayyar, "An Initial Analysis of 85 Kr Production and Dispersion from Reprocessing in India and Pakistan," *Science & Global Security* 10 (2002): 151–179.

104. See for instance, "The President's News Conference, November 30, 1950," Public Papers of the Presidents: Harry S. Truman, 1945–1953, www.trumanlibrary.org. For other instances, see Roger Dingman, "Atomic Diplomacy during the Korean War," *International Security* 13, no. 3 (1989): 50–91. See also McGeorge Bundy, *Danger and Survival: Choices about the Bomb in the First Fifty Years* (New York: Random House, 1988), 231–245.

105. Glenn Kessler, "Message to U.S. Preceded Nuclear Declaration by North Korea," *Washington Post*, July 2, 2008; "North Korea Declares 31 Kilograms of Plutonium," *Global Security Newswire*, October 24, 2008.

106. David Albright, "How Much Plutonium Did North Korea Produce?," in David Albright and Kevin O'Neill, eds., *Solving the North Korean Nuclear Puzzle* (Washington, DC: Institute for Science and International Security, 2000), 111–126. See also *Global Fissile Material Report 2009: A Path to Nuclear Disarmament* (Princeton, NJ: International Panel on Fissile Materials, October 2009), 48–51, www.fissilematerials.org/library/gfmr09.pdf.

107. Glenn Kessler, "Far-Reaching U.S. Plan Impaired N. Korea Deal," *Washington Post*, September 26, 2008.

108. North Korea had about 50 tons of spent fuel rods from its Yongbyon reactor, which could have contained about 10 kilograms of plutonium (*Global Fissile Material Report 2009*, 51).

109. This estimate assumes that North Korea consumed about 5 kilograms of plutonium in each of its nuclear tests, in 2006 and 2009, out of the roughly 30–40 kilograms of plutonium it had declared to have been produced by 2008. There have been suggestions of an undeclared test in 2010; see Lars-Erik De Geer, "Radionuclide Evidence for Low-Yield Nuclear Testing in North Korea in April/May 2010," *Science & Global Security* 20 (2012): 1–29; David P. Schaff, Won-Young Kim, and Paul G. Richards, "Seismological Constraints on Proposed Low-Yield Nuclear Testing in Particular Regions and Time Periods in the Past," *Science & Global Security* 20 (2012): 155–171; Christopher M. Wright, "Low-Yield Nuclear Testing by North Korea in May 2010: Assessing the Evidence with Atmospheric Transport Models and Xenon Activity Calculations," *Science & Global Security* 21 (2013): 3–52. North Korea carried out an announced test in 2013. It is not known whether the possible test in 2010 or the declared test in 2013 used plutonium or highly enriched uranium.

110. Pakistan's former president General Pervez Musharraf has revealed that A. Q. Khan transferred nearly two dozen P-1 and P-2 centrifuges to North Korea. Khan also "provided North Korea with a flow meter, some special oils for centrifuges, and coaching on centrifuge technology, including visits to top-secret centrifuge plants." Pervez Musharraf, *In the Line of Fire: A Memoir* (New York: Free Press, 2006), 296.

111. The 2002 CIA report and other U.S. assessments are described in Mary Beth Nikitin, *North Korea's Nuclear Weapons: Technical Issues*, RL34256 (Washington, DC: Congressional Research Service, April 3, 2013).

112. Siegfried S. Hecker, *A Return Trip to North Korea's Yongbyon Nuclear Complex* (Center for International Security and Cooperation, Stanford University, November 20, 2010); Peter Crail, "N. Korea Reveals Uranium Enrichment Plant," *Arms Control Today*, December 2010.

113. Belarus, Kazakhstan, and the Ukraine inherited nuclear weapons when the Soviet Union disintegrated in 1991 but agreed to transfer them to Russia.

114. Gavin Ball, "Status of Conversion of the South African Safari-I Reactor and ^{99}Mo Production Process to Low Enriched Uranium" (presented at the 32nd International Meeting on Reduced Enrichment for Research and Test Reactors, Lisbon, 2010). The United States also provided slightly enriched uranium and heavy water for a critical assembly mockup of a heavy water reactor, Safari-II, which South Africa was planning to use for research on plutonium production and separation.

115. Waldo Stumpf, "South Africa's Nuclear Weapons Program: From Deterrence to Dismantlement," *Arms Control Today* (December 1995): 3–8.

116. A. J. A. Roux et al., "Development and Progress of the South African Enrichment Project," in *Proceedings of the International Conference on Nuclear*

Power and Its Fuel Cycles; Salzburg, Austria; 2–13 May 1977, 182 (Vienna, Austria: International Atomic Energy Agency, 1977), http://www.iaea.org/inis/collection/NCLCollectionStore/_Public/08/303/8303321.pdf.

117. The eastern section of the Pelindaba Site, where the enrichment plants were located, is also called Valindaba.

118. Stumpf, "South Africa's Nuclear Weapons Program." For a timeline of South Africa's weapons program, see Adolf von Baeckmann, Garry Dillon, and Demetrius Perricos, "Nuclear Verification in South Africa," *IAEA Bulletin* 37, no. 1 (1995): 45.

119. In April 1981, South Africa publicly announced that it "was now producing a limited quantity of 45 per cent enriched uranium"; a statement that was "bound to rekindle international fears about the country's nuclear intentions." Nicholas Ashford, "South Africans Now Able to Produce A-Bomb," *Times* (London), April 30, 1981.

120. At the time, South Africa did not yet have sufficient HEU from domestic production available for a nuclear explosive. A cold nuclear test would have used natural uranium in lieu of HEU to test the mechanical performance of the design in a configuration that later could have been used for an actual nuclear test. There have also been speculations that the site was prepared for a clandestine Israeli test, but there is no evidence to support this theory. At the time, South Africa and Israel collaborated closely on military and nuclear affairs.

121. The detection of these preparations by reconnaissance satellites was not accidental as a high-level South African official, Dieter Gerhardt, had been passing information to Moscow for years. Polakow-Suransky, *The Unspoken Alliance*, 112.

122. The production of 50–60 kilograms of weapon-grade HEU per year would correspond to a capacity of the Y-Plant on the order of 10,000–12,000 SWU/yr.

123. On the decisions to dismantle the South African nuclear weapons program, see, for example, Peter Liberman, "The Rise and Fall of the South African Bomb," *International Security* 26, no. 2 (Fall 2001): 45–86.

124. Von Baeckmann, Dillon, and Perricos, "Nuclear Verification in South Africa."

125. Six weapons were fully assembled and an additional weapon was under construction when the program was abandoned. See ibid., 45. Among the six assembled weapons, five were air-deliverable and one would have been used in a nuclear test to "reveal" South Africa's capability.

126. Thomas B. Cochran, "Highly Enriched Uranium Production for South African Nuclear Weapons," *Science & Global Security* 4, no. 1 (1994): 161–176.

127. South Africa's HEU stockpile has been used to make fuel for the Safari reactor and irradiation targets for the production of medical radioisotopes.

128. Von Baeckmann, Dillon, and Perricos, "Nuclear Verification in South Africa," 48.

4   The Global Stockpile of Fissile Material

1. *Historical Accounting for UK Defense Highly Enriched Uranium* (UK Ministry of Defense, March 2006), www.fissilematerials.org/library/mod06.pdf.

2. *Declassification of Today's Highly Enriched Uranium Inventories at Department of Energy Laboratories* (Washington, DC: U.S. Department of Energy, Office of the Press Secretary, June 27, 1994), www.fissilematerials.org/library/doe06a.pdf.

3. For a general discussion of the political and technical dimensions of nuclear transparency, see Nicholas Zarimpas, *Transparency in Nuclear Warheads and Materials: The Political and Technical Dimensions*, Stockholm International Peace Research Institute (Oxford: Oxford University Press, 2003).

4. For the "Action Plan on Nuclear Disarmament" see *2010 Review Conference of the Parties to the Treaty on the Non-Proliferation of Nuclear Weapons, Final Document, Volume 1* (New York: United Nations, 2010), 19–21, www.un.org/en/conf/npt/2010.

5. INFCIRC denotes an information circular published by the IAEA that records an agreement by a state with or report made to the IAEA.

6. The 1998 Strategic Defense Review noted: "All stocks of highly enriched uranium will . . . be retained outside safeguards, since material no longer needed for nuclear weapons will be used for the naval propulsion programme." Supporting Essay Five: Deterrence, Arms Control, and Proliferation, §26, *The Strategic Defense Review*, Cm 3999 (UK Ministry of Defense, July 1998), www.fissilematerials.org/library/mod98.pdf.

7. All U.S., French, British, most Russian, and some Chinese submarines are nuclear powered, as are all U.S. aircraft carriers. Brazil and reportedly also Iran are pursuing nuclear-powered submarine projects. At the same time, however, both France and the United Kingdom have decided for cost reasons to use conventional power plants for their future aircraft carriers, and several countries have developed advanced conventional submarines.

8. For example, if the fresh HEU is 90 percent enriched and assuming that half of the uranium-235 is consumed during irradiation, then the spent fuel would still be more than 70 percent enriched.

9. *Global Fissile Material Report 2010: Balancing the Books: Production and Stocks*, chap. 4 (Princeton, NJ: International Panel on Fissile Materials, 2010), www.fissilematerials.org/library/gfmr10.pdf. Some Russian spent naval fuel is currently in indefinite storage but the intention is to reprocess it. In addition to reprocessing and storage, the former Soviet Union also dumped into the Arctic seas fourteen nuclear reactors and a nuclear submarine (K-27) with its two HEU-fueled (lead-bismuth cooled) reactors. Charles Digges, "Russia Announces Enormous Finds of Radioactive Waste and Nuclear Reactors in Arctic Seas," *Bellona*, August 28, 2012, www.bellona.org/articles/articles_2012/Russia_reveals_dumps.

10. This is an interpolation between 14 tons in 1997 and an estimated 65 tons in 2035. David Curtis, "Naval Nuclear Propulsion Program; Introduction to Spent Naval Fuel" (presented at the U.S. Nuclear Waste Technical Review Board, Meeting of the Panel on the Repository, Augusta, GA, December 17, 1997), www.nwtrb.gov/meetings/1997/dec/curtis.pdf.

11. The IAEA publishes the number of significant quantities (SQs) under safeguards in its Annual Reports (Table A4 in the Annex of the IAEA reports). At the end of 2012, 211 SQs of HEU were under comprehensive safeguards agreements in NPT non-weapon states. This corresponds to 5.3 tons of uranium-235 contained in HEU; the amount of HEU is higher and depends on the average enrichment level, which is not made public. Germany is the only non-weapon state that annually submits public reports of its HEU stocks to the IAEA. As with France and the United Kingdom, this is done in an appendix to its INFCIRC/549 report on its plutonium stocks.

12. See the twenty-five-year overview of the Reduced Enrichment for Research and Test Reactor program in Armando Travelli, "Status and Progress of the RERTR Program in the Year 2003" (presented at the 2003 International Meeting on Reduced Enrichment for Research and Test Reactors, Chicago, 2003).

13. In April 2009, President Obama announced the goal of securing "all vulnerable nuclear material around the world within four years" and launched a series of global Nuclear Security Summits that would be held in Washington, DC (2010), Seoul (2012), The Hague (2014), and Washington, DC (2016). One major focus of these summits has been research reactor conversion to low-enriched fuel. U.S. funding to support these efforts had already been increasing dramatically since the creation of the Global Threat Reduction Initiative (GTRI) in 2004. The GTRI enjoys broad international support.

14. Agreement between the Government of the United States of America and the Government of the Russian Federation Concerning the Disposition of Highly Enriched Uranium Extracted from Nuclear Weapons, February 18, 1993. www .fissilematerials.org/library/heu93.pdf.

15. "Final Megatons to Megawatts Shipment Completes Historic Program," USEC, December 10, 2013, www.usec.com. USEC is the U.S. company that purchased from Russia the LEU made by down blending 500 tons of HEU.

16. *Amended Record of Decision: Disposition of Surplus Highly Enriched Uranium Environmental Impact Statement* (Washington, DC: National Nuclear Security Administration, U.S. Department of Energy, April 29, 2011), 7–8.

17. Alexander Glaser and M. V. Ramana, "Weapon-Grade Plutonium Production Potential in the Indian Prototype Fast Breeder Reactor," *Science & Global Security* 15, no. 2 (2007): 85–105.

18. The 53.7 tons of excess U.S. plutonium cited here does not include material that has been declared excess but is still contained in spent fuel. *The United States Plutonium Balance, 1944–2009* (Washington, DC: U.S. Department of Energy, June 2012).

19. A set of proposals for increasing transparency of fissile material and nuclear warhead stocks is offered in *Global Fissile Material Report 2013: Increasing*

Transparency of Nuclear Warhead and Fissile Material Stocks as a Step toward Disarmament (Princeton, NJ: International Panel on Fissile Materials, November 2013), www.fissilematerials.org/library/gfmr13.pdf.

20. The UK declaration noted: "This review has been conducted from an audit of annual accounts and the delivery/receipt records at sites. A major problem encountered in examining the records was that a considerable number had been destroyed from the early years of the programme . . . Even where records have survived, other problems have been encountered, including . . . distinction between new make and recycled HEU . . . some early records make no specific mention of waste and effluent disposals . . . [for] some records . . . assessments had to be made to establish units. Other records do not identify quantities to decimal places and . . . may have been rounded . . . [and] in some cases no indication of enrichment value was available." *Historical Accounting for UK Defence Highly Enriched Uranium*, 2.

21. "Marcoule: Dismantling the G1, G2 and G3 Reactors" (Commissariat à l'Energie Atomique, 2009), www.francetnp.fr; "Marcoule: Dismantling the UP1 Reprocessing Plant" (Commissariat à l'Energie Atomique, 2009), www .francetnp.fr.

22. For a review of several case studies, see T. W. Wood et al., "Establishing Confident Accounting for Russian Weapons Plutonium," *Nonproliferation Review* 9, no. 2 (Summer 2002): 126–137. Equivalent methods have been proposed for other types of reactors, especially for heavy water-moderated reactors. Alex Gasner and Alexander Glaser, "Nuclear Archaeology for Heavy-Water-Moderated Plutonium Production Reactors," *Science & Global Security* 19, no. 3 (2011): 223–233. This technique was first suggested in Steve Fetter, "Nuclear Archaeology: Verifying Declarations of Fissile-Material Production," *Science & Global Security* 3, nos. 3–4 (1993).

5   Fissile Materials, Nuclear Power, and Nuclear Proliferation

1. "The Statute of the IAEA" (International Atomic Energy Agency, Vienna, 1956), www.iaea.org/About/statute.html.

2. Mohamed ElBaradei, "Towards a Safer World," *The Economist*, October 16, 2003.

3. The U.S. Atomic Energy Act of 1946 prohibited any nuclear cooperation until Congress could establish that effective safeguards against nuclear weapon proliferation were in place. It cut off sharing even with the United Kingdom. This was despite the September 1944 Hyde Park agreement between President Roosevelt and Prime Minister Churchill that "full collaboration" in nuclear development "for military and commercial purposes should continue after the defeat of Japan unless and until terminated by joint agreement." *Aide Memoire of Conversation Between the President and the Prime Minister at Hyde Park, September 18, 1944, in U.S. Department of State, Foreign Relations of the United States: Conference at Quebec, 1944* (Washington, DC: U.S. Government Printing Office, 1944), 492–493. In June 1943, General Leslie Groves and the Military Policy

Committee, which managed the Manhattan Project, adopted a goal that the United States should seek as complete control as possible of the world uranium supply. Richard G. Hewlett and Oscar E. Anderson, *The New World: A History of the United States Atomic Energy Commission, Volume 1, 1939–1946*, California Studies in the History of Science (University of California Press, 1962). The June 1945 Franck Report had argued that seeking to monopolize the uranium supply was not a viable option since significant reserves were likely to be found in territory not controlled by the United States and its allies. James Franck et al., *Report of the Committee on Political and Social Problems Manhattan Project* (The Franck Report) (University of Chicago, June 11, 1945), www.fissilematerials.org/library/fra45.pdf.

4. Dwight D. Eisenhower, "Peaceful Uses of Atomic Energy" (presented at the General Assembly of the United Nations, New York, December 8, 1953).

5. "Preliminary Proposal for an International Organization to Further the Uses of Atomic Energy," June 8, 1954; C. D. Jackson private papers, Box 29, Atomic Industrial Forum, Eisenhower Presidential Museum and Library.

6. *Difficulties in Determining if Nuclear Training of Foreigners Contributes to Weapons Proliferation, Report by the Comptroller General of the United States* (Washington, DC: General Accounting Office, April 23, 1979), 18, 20.

7. Between 1955 and 1977, 13,456 foreigners from eighty-four countries participated in research at U.S. nuclear facilities. Ibid., 83.

8. Ibid., 79.

9. George Perkovich, "Nuclear Power and Nuclear Weapons in India, Pakistan, and Iran," in Paul L. Leventhal, Sharon Tanzer, and Steven Dolley, eds., *Nuclear Power and the Spread of Nuclear Weapons: Can We Have One without the Other?* (Washington, DC: Brassey's, 2002), 194.

10. Bertrand Goldschmidt, *The Atomic Complex: A Worldwide Political History of Nuclear Energy* (La Grange Park, IL: American Nuclear Society, 1982), 259–261. On the papers published on plutonium separation techniques, see "Chemical Processing of Irradiated Fuel Elements "(Volume 9, Session 21B) in *Proceedings of the International Conference on the Peaceful Uses of Atomic Energy Held in Geneva, August 8–20, 1955* (New York: United Nations, 1956).

11. R. Scott Kemp, "Nonproliferation Strategy in the Centrifuge Age" (PhD thesis, Princeton University, 2010), chap. 3.

12. Gernot Zippe, *The Development of the Short Bowl Ultracentrifuge*, University of Virginia, Oak Ridge Operations Report, ORO-315 (U.S. Atomic Energy Commission, July 1960).

13. These nine countries are Australia, Brazil, China, India, Israel, Italy, Japan, France, and Sweden. Kemp, "Nonproliferation Strategy in the Centrifuge Age."

14. In the mid-1980s, A.Q. Khan, having set up Pakistan's enrichment program, began to sell centrifuge designs, equipment, and expertise to Iran, Libya, North Korea, and perhaps others. In the case of Libya, this included making a deal to supply an entire enrichment plant.

15. International Atomic Energy Agency, Nuclear Fuel Cycle Information System (NFCIS), infcis.iaea.org.

16. Countries that at one time had nuclear weapons programs or considered the acquisition of nuclear weapons but did not acquire them include: Algeria, Argentina, Australia, Brazil, Canada, Egypt, Germany, Indonesia, Iraq, Italy, Japan, Libya, South Korea, Sweden, Switzerland, and Taiwan. Ariel Levite, "Never Say Never Again: Nuclear Reversal Revisited," *International Security* 27, no. 3 (Winter 2002).

17. The President's News Conference, March 21, 1963, www.presidency.ucsb.edu.

18. *Treaty on the Non-Proliferation of Nuclear Weapons*. INFCIRC/140. Vienna: International Atomic Energy Agency, April 22, 1970, Art. IX.

19. United Nations Office of Disarmament Affairs, *Treaty on the Non-Proliferation of Nuclear Weapons, Status of the Treaty*, www.un.org/disarmament.

20. These countries joined the NPT in 1993/94 after the breakup of the Soviet Union, from which they inherited large stockpiles of nuclear weapons over which they had, however, little or no control. In total, over 6,000 strategic and tactical nuclear warheads were removed from these countries between 1991 and 1996.

21. *Treaty on the Non-Proliferation of Nuclear Weapons*. INFCIRC/140. Vienna: International Atomic Energy Agency, April 22, 1970.

22. *The Structure and Content of Agreements between the Agency and States Required in Connection with the Treaty on the Non-Proliferation of Nuclear Weapons*, INFCIRC/153 (Corrected) (Vienna: International Atomic Energy Agency, June 1972).

23. International Atomic Energy Agency, www.iaea.org/safeguards/what.html.

24. *IAEA Annual Report 2012* (Vienna: International Atomic Energy Agency, 2013). The number of facilities under safeguards is given in the Annex of the IAEA Report, Table A5. There are 1,285 facilities under safeguards in non-weapon states. The cost of safeguards, including the regular budget and extra-budgetary contributions by states, is given in Tables A1 and A2 of the IAEA Report.

25. *IAEA Safeguards Glossary, 2001 Edition* (Vienna: International Atomic Energy Agency, 2003), 19–29.

26. *Model Protocol Additional to the Agreement(s) Between State(s) and the International Atomic Energy Agency for the Application of Safeguards*, INFCIRC/540 (Corrected) (Vienna: International Atomic Energy Agency, September 2007).

27. International Atomic Energy Agency, *Conclusion of Additional Protocols: Status as of 12 March 2014*, www.iaea.org/safeguards/documents/AP_status_list.pdf.

28. J. Samuel Walker and George T. Mazuzan, *Containing the Atom: Nuclear Regulation in a Changing Environment, 1963–1971* (Berkeley: University of California Press, 1992), 227–230.

29. *Report to the Atomic Energy Commission by the Ad Hoc Advisory Panel on Safeguarding Special Nuclear Material*, March 10, 1967, 4.

30. Mason Willrich and Theodore B. Taylor, *Nuclear Theft: Risks and Safeguards* (Cambridge, MA: Ballinger, 1974); John McPhee, *The Curve of Binding Energy* (New York: Farrar, Straus and Giroux, 1974).

31. Fears about the risk of theft of fissile materials led to the Convention on the Physical Protection of Nuclear Material, established in March 1980, as a means to set common security standards for nuclear materials in international transport. In July 2005, the Convention was renamed the Convention on Physical Protection of Nuclear Material and Nuclear Facilities and amended to add an explicit obligation to protect nuclear materials in domestic use, storage, and transport. The amendments will take effect once they have been ratified by two-thirds of the States Parties to the Convention. As of the end of 2013, there were 149 parties to the Convention and 74 states had agreed to the amendment. International Atomic Energy Agency, *Status of the Convention on Physical Protection of Nuclear Material (CPPNM) and Its Amendment, as of 17 December 2013 and 27 March 2014, respectively.*

32. Note that an incident is included in the database only if formally reported or "confirmed" by a participating state. In 2007, there were about 800 additional known "unconfirmed" incidents not included in the database. *International Conference on Illicit Nuclear Trafficking: Collective Experience and the Way Forward*, IAEA-CN-154 (International Atomic Energy Agency, November 2007).

33. IAEA Incident and Trafficking Database (ITDB): Incidents of Nuclear and Other Radioactive Material Out of Regulatory Control 2013, Fact Sheet, International Atomic Energy Agency, Vienna, 2013.

34. United Nations Security Resolution S/RES/1540 (2004), April 28, 2004.

35. "Letter dated 27 December 2012 from the Chair of the Security Council Committee established pursuant to resolution 1540 (2004) addressed to the President of the Security Council," UN Security Council S/2012/963, December 28, 2012.

36. The 2012 report from the chair of the 1540 Committee to the United Nations Security Council noted that "further efforts are required in the area of implementation . . . [and] more intensive efforts are needed to encourage the formal submission of voluntary additional information by those States that have already submitted reports. . . . Despite their commitment to the implementation of resolution 1540 (2004), some States lack the required legal and regulatory infrastructure, implementation experiences and/or resources to do so." Letter dated December 27, 2012.

37. IAEA, Power Reactor Information System, pris.iaea.org/public. The thirty countries include Iran, where the Bushehr reactor came online in 2011.

38. The literature on this subject is extensive; for an excellent perspective, see John P. Holdren, "Civilian Nuclear Technologies and Nuclear Weapons

Proliferation," in Carlo Schaerf, Brian Holden Reid, and David Carlton, eds., *New Technologies and the Arms Race* (London: Macmillan, 1989), 161–198.

39. Harold A. Feiveson, "Latent Proliferation: The International Security Implications of Civilian Nuclear Power" (PhD thesis, Princeton University, 1972).

40. J. Carson Mark, "Explosive Properties of Reactor-Grade Plutonium," *Science & Global Security* 4, no. 1 (1993): 111–128.

41. Houston Wood, Alexander Glaser, and R. Scott Kemp, "The Gas Centrifuge and Nuclear Weapons Proliferation," *Physics Today* (September 2008): 40–45.

42. It is worth noting that the IAEA was able to detect 20 percent enriched uranium in Iran's pilot enrichment plant shortly after Iran had announced its plans to begin production of this material. *Implementation of the NPT Safeguards Agreement and relevant provisions of Security Council resolutions 1737 (2006), 1747 (2007), 1803 (2008) and 1835 (2008) in the Islamic Republic of Iran*, International Atomic Energy Agency, GOV/2010/28, Vienna, May 31, 2010.

43. Alexander Glaser, "Characteristics of the Gas Centrifuge for Uranium Enrichment and Their Relevance for Nuclear Weapon Proliferation," *Science & Global Security* 16, nos. 1–2 (2008): 1–25.

44. D. E. Ferguson, "Simple, Quick Processing Plant," Intra-Laboratory Correspondence (Oak Ridge National Laboratory, August 30, 1977); and *Quick and Secret Construction of Plutonium Reprocessing Plants: A Way to Nuclear Weapons Proliferation?* EMD-78–104 (Report to the Comptroller General of the United States, October 6, 1978). Similar conclusions were reached in subsequent U.S. assessments: J. P. Hinton et al., *Proliferation Resistance of Fissile Material Disposition Program Plutonium Disposition Alternatives: Report of the Proliferation Vulnerability Red Team*, SAND97–8201 (Sandia National Laboratory, October 1996), www.fissilematerials.org/library/doe96c.pdf; and Victor Gilinsky, Marvin Miller, and Harmon Hubbard, *A Fresh Examination of the Proliferation Dangers of Light Water Reactors* (Washington, DC: The Nonproliferation Policy Education Center, October 22, 2004).

45. An enrichment plant with 5,000 first-generation centrifuges could make enough HEU for one bomb a year. It would require a floor area approximately 50 meters on a side, easily able to fit in a small building or underground, and would consume only about 100 kilowatts of electrical power, which could be provided by a diesel generator. Plants of this kind were built by Pakistan in the late 1970s and early 1980s and more recently in Iran. *Global Fissile Material Report 2007* (Princeton, NJ: International Panel on Fissile Materials, September 2007), chap. 9. See also R. S. Kemp and A. Glaser, "The Gas Centrifuge and the Nonproliferation of Nuclear Weapons," in Shi Zeng, ed., *Proceedings of the Ninth International Workshop on Separation Phenomena in Liquids and Gases, 18–21 September 2006* (Beijing: Tsinghua University Press, 2007), 88–95.

46. Kemp, "Nonproliferation Strategy in the Centrifuge Age."

47. For a more extended discussion, see *Global Fissile Material Report 2009: A Path to Nuclear Disarmament* (Princeton, NJ: International Panel on Fissile Materials, October 2009), chap. 8, www.fissilematerials.org/library/gfmr09.pdf.

48. See Tadahiro Katsuta and Tatsujiro Suzuki, *Japan's Spent Fuel and Pluto-nium Management Challenges* (International Panel on Fissile Materials, September 2006); Frank von Hippel, *Managing Spent Fuel in the United States: The Illogic of Reprocessing* (International Panel on Fissile Materials, January 2007); Mycle Schneider and Yves Marignac, *Spent Nuclear Fuel Reprocessing in France* (International Panel on Fissile Materials, April 2008); Martin Forwood, *The Legacy of Reprocessing in the United Kingdom* (International Panel on Fissile Materials, July 2008). All these reports are available at www.fissilematerials.org.

49. Yuri Yudin, *Multilateralization of the Nuclear Fuel Cycle: Assessing the Existing Proposals* (Geneva: United Nations Institute for Disarmament Research, 2009), www.unidir.ch; Alexander Glaser, *Internationalization of the Nuclear Fuel Cycle*, ICNND Research Paper No. 9 (International Commission on Nuclear Non-proliferation and Disarmament, February 2009), www.icnnd.org.

50. Germany has proposed a Multilateral Enrichment Sanctuary Project, a scheme that envisages a host state offering a site on its territory to a separate set of countries to build and operate an enrichment plant on that site. Ideally, the host would have no experience with uranium enrichment so that a hypothetical takeover of the plant would be less of a concern. *Communication Received from the Resident Representative of Germany to the IAEA with Regard to the German Proposal on the Multilateralization of the Nuclear Fuel Cycle*, IAEA, INFCIRC/704, May 4, 2007. It was later amended in INFCIRC/727, May 30, 2008, and INFCIRC/735, September 25, 2008. For a discussion, see Glaser, *Internationalization of the Nuclear Fuel Cycle*.

51. International ownership of enrichment and reprocessing plants was proposed in the 1946 Acheson-Lilienthal Report. Similar ideas were discussed in the 1970s when the creation of an International Nuclear Fuel Authority (INFA) was considered by the United States. More recently, some analysts have picked up the INFA concept as a strategy to resolve the crisis surrounding Iran's enrichment program: Thomas B. Cochran and C. E. Paine, "International Management of Uranium Enrichment" (presented at the International Meeting on Nuclear Energy and Proliferation in the Middle East, Amman, Jordan, June 22, 2009).

52. Robert Socolow and Alexander Glaser, "Balancing Risks: Nuclear Energy and Climate Change," *Daedalus* 138, no. 4 (2009): 31–44.

6   Ending the Separation of Plutonium

1. The 3 grams of uranium-238 in an average ton of crustal rock has a releasable energy of about 3 megawatt-days (thermal) or 260×10^9 joules, which is about ten times the energy in a ton of coal.

2. Leo Szilard, "Liquid Metal Cooled Fast Neutron Breeders," March 6, 1945, reprinted in Bernard T. Feld et al., *The Collected Works of Leo Szilard* (Cambridge, MA: MIT Press, 1972), 369. Breeder reactors were invoked in the earliest post–World War II papers on nuclear power; for example, "A mother pile [reactor] would supply the distant power plant with fissionable material,"

J. Marschak, "The Economics of Atomic Power," *Bulletin of the Atomic Scientists of Chicago*, February 15, 1946, 3.

3. Alfred Perry and Alvin Weinberg, "Thermal Breeder Reactors," *Annual Review of Nuclear Science* 22 (1972): 317. As discussed in chapter 2, uranium-233 is a fissile material with about the same critical mass as plutonium. It is bred in a reactor by absorption of a neutron in natural thorium (thorium-232) and subsequent radioactive decay of the thorium-233.

4. See, for example, Glenn T. Seaborg, "The Plutonium Economy of the Future" (presented at the Fourth International Conference on Plutonium and Other Actinides, Santa Fe, New Mexico, 1970).

5. Richard Garwin, "The Role of the Breeder Reactor," in Frank Barnaby, ed., *Nuclear Energy and Nuclear Weapon Proliferation* (London: Taylor and Francis, 1979), 141.

6. The proton nuclei of the hydrogen atoms in water have approximately the same mass as a neutron. A neutron can therefore lose up to all its energy in a single collision with a proton—just as one billiard ball does in a frontal collision with another.

7. Leo Szilard, "Liquid Metal Cooled Fast Neutron Breeders."

8. Richard G. Hewlett and Francis Duncan, *Nuclear Navy, 1946–1962* (Chicago: University of Chicago Press, 1974), 274.

9. Thomas B. Cochran et al., *Fast Breeder Reactor Programs: History and Status* (Princeton, NJ: International Panel on Fissile Materials, February 2010), 6, www.fissilematerials.org/library/rr08.pdf.

10. Cochran et al., *Fast Breeder Reactor Programs.*

11. O. M. Saraev, "Operating Experience with Beloyarsk Fast Reactor BN600 [Nuclear Power Plant]," *Unusual Occurrences During LMFR Operation, Proceedings of a Technical Committee meeting held in Vienna, 9–13 November 1998*, IAEA-TECDOC-1180 (Vienna: International Atomic Energy Agency: October 2000), 101–116. There were thirteen other leaks of sodium into the environment and twelve leaks in the steam generators during the same period. However, no leaks were reported during the following thirteen years. N. N. Oshkanov et al., "30 Years of Experience in Operating the BN-600 Sodium-Cooled Fast Reactor," *Atomic Energy* 108, no. 4 (2010): 234–239.

12. R. B. Fitts and H. Fujii, "Fuel Cycle Demand, Supply and Cost Trends," *IAEA Bulletin* 18, no. 1 (1976): 19. The recovery cost range considered was up to $30 per pound of U_3O_8 (1975 $). This corresponds to about $260 per kilogram in 2010 $. We assume uranium requirements for a once-through fuel cycle of 160–200 tons per GWe-year.

13. Table 3 in *Energy, Electricity and Nuclear Power Estimates for the Period up to 2050,* (Vienna: International Atomic Energy Agency, 2013).

14. Identified plus estimated undiscovered resources at estimated recovery costs up to $130/kgU, Organization for Economic Cooperation and Development Nuclear Energy Agency and International Atomic Energy Agency, *Uranium 2011:*

Resources, Production and Demand (OECD Nuclear Energy Agency, 2012), Tables 1.10, 1.11, 1.14.

15. The cost of nuclear power for the first decades of operation of a nuclear power reactor is dominated by the capital cost of the reactor. After power reactors have been paid for, operating them typically costs only a few cents per kWh.

16. See, for example, Erich A. Schneider and William C. Sailor, "Long-term Uranium Supply Estimates," *Nuclear Technology* 162, no. 3 (2008): 379–387.

17. France's Radioactive Materials and Waste Planning Act of 2006, as summarized in English by the national radioactive waste agency, ANDRA, requires (Article 3.1) "an assessment of the industrial prospects of [transmutation] systems and to commission a pilot facility before 31 December 2020," www.andra.fr. France's nuclear energy establishment has interpreted this as a requirement to build another fast-neutron sodium-cooled reactor.

18. The breeder designs of the 1970s required a startup inventory of 5 to 6 tons of fissile plutonium (plutonium-239 and plutonium-241) or about 7 to 10 tons of total light-water reactor plutonium per GW(e), including one annual reload. Table II in *Fast Breeders, International Nuclear Fuel Cycle Evaluation (INFCE)* (Vienna: International Atomic Energy Agency, 1980).

19. *A Brief History of Reprocessing and Cleanup in West Valley, NY, Fact Sheet* (Union of Concerned Scientists, December 2007).

20. Anthony Andrews, *Nuclear Fuel Reprocessing: U.S. Policy Development*, RS22542 (Washington, DC: Congressional Research Service, Library of Congress, March 27, 2008); J. Michael Martinez, "The Carter Administration and the Evolution of American Nuclear Nonproliferation Policy, 1977–1981," *Journal of Policy History* 14, no. 3 (2002) 261–292.

21. Andrews, *Nuclear Fuel Reprocessing*, 3.

22. Ibid.

23. *Comparative Analysis of Alternative Financing Plans for the Clinch River Breeder Reactor Project* (Washington, DC: U.S. Congressional Budget Office, 1983).

24. Cochran et al., *Fast Breeder Reactor Programs*, 103.

25. In addition, India has a fourth reprocessing plant at Trombay, which has been used to separate plutonium for weapons.

26. Harold A. Feiveson et al., *Managing Spent Fuel from Power Reactors: Experience and Lessons from Around the World* (Princeton, NJ: International Panel on Fissile Materials, September 2011).

27. *Oxide Fuels, Preferred Option* (UK Nuclear Decommissioning Authority, June 2012), www.nda.gov.uk.

28. Heavy water reactor (HWR) spent fuel contains about 0.35 percent. See the appendix in *Global Fissile Material Report 2010: Balancing the Books: Production and Stocks* (Princeton, NJ: International Panel on Fissile Materials, 2010), www.fissilematerials.org/library/gfmr10.pdf. UK advanced gas-cooled reactors (AGR) spent fuel contains about 0.55 percent plutonium and Magnox fuel about

0.3 percent. David Albright, Frans Berkhout, and William Walker, *Plutonium and Highly Enriched Uranium 1996: World Inventories, Capabilities and Policies* (Oxford: Oxford University, 1997), 136–138.

29. Mitchell Reiss, *Bridled Ambition: Why Countries Constrain Their Nuclear Capabilities* (Baltimore, MD: Johns Hopkins University Press, 1995); Kurt M. Campbell, Robert J. Einhorn, and Mitchell Reiss, *The Nuclear Tipping Point: Why States Reconsider Their Nuclear Choices* (Washington, DC: Brookings Institution Press, 2004); Thomas Johansson, "Sweden's Abortive Nuclear Weapons Project," *Bulletin of the Atomic Scientists*, March 1986.

30. "Tokai Plutonium Removal Figures Revised," *Japan Times*, April 2, 2003. The IAEA has a second line of defense, "containment and surveillance," which is seals and motion-activated cameras to detect any unusual activity at a potential diversion point. However, seals, at least, are regularly broken by accident and the only recourse then is to go back to uncertain measurements to see whether any material is missing.

31. Plutonium dioxide powder is exceedingly carcinogenic if inhaled. The inhalation of total one gram by a million people would result in on the order of 10,000 extra cancer deaths. Steve Fetter and Frank von Hippel, "The Hazard from Plutonium Dispersal by Nuclear-Warhead Accidents," *Science & Global Security* 2 (1990).

32. W. R. Lloyd, *Dose Rate Estimates from Irradiated Light-Water-Reactor Fuel Assemblies in Air*, UCRL-ID-115199 (Lawrence Livermore National Laboratory, 1994). For fuel with a fission energy release of 30 MWt-days/kgU, the dose rate at fifty years is about 9.2 Grays per hour (Gy/hr). For BWR fuel, it is about 4.7 Gy/hr. After about a hundred years, the dose rate drops below 1 Gy/hr, and the spent fuel is no longer considered "self-protecting." In a comprehensive assessment after the Chernobyl accident, it was concluded that the $LD_{50/60}$ dose—that is, the dose for which half of those exposed would die of radiation illness within sixty days—is 2.5 Gy without treatment, 5 Gy "with good supportive medical treatment," and "probably up to around 9 Gy" with "successful marrow transplantation." *Sources and Effects of Ionizing Radiation* (New York: United Nations Scientific Committee on the Effects of Atomic Radiation [UNSCEAR], Report to the General Assembly, 1988), 596.

33. A glove box is an enclosed volume, usually made in large part of transparent plastic and kept below atmospheric pressure, with sleeves and gloves built in to allow workers to process toxic materials inside without being exposed to an inhalation hazard.

34. The percentage of plutonium is higher than that of uranium-235 because the chain-reacting plutonium isotopes come mixed with the plutonium isotopes that do not chain-react with slow neutrons: plutonium-238, plutonium-240, and plutonium-242.

35. Plutonium from LEU fuel with a burn-up of 43 MWt-days per kilogram heavy metal contains 53 percent plutonium-239. Plutonium from MOX spent fuel made with the above plutonium contains 37 percent plutonium-239. The fraction of plutonium-241 is slightly larger: 15 percent for the LEU fuel and 17

percent for the MOX fuel, but the plutonium-241 decays away with fourteen-year half-life. Tables 9 and 12 in *Plutonium Fuel: An Assessment, Report by an Expert Group* (Paris: OECD Nuclear Energy Agency, 1989).

36. The percentage of uranium-235 in the recovered uranium is about the same as in natural uranium. It also contains uranium-236, however, which absorbs neutrons without fissioning. To overcome the neutron-absorbing effect of the uranium-236, it is necessary to increase the uranium enrichment by about half a percent. Figure 7 in *Management of Reprocessed Uranium: Current Status and Future Prospects*, IAEA-TECDOC-1529 (Vienna: International Atomic Energy Agency, February 2007).

37. J. M. Charpin, B. Dessus, and R. Pellat, *Economic Forecast Study of the Nuclear Power Option, Report to the Prime Minister*, July 2000. Tables on pp. 43, 56, 214, and 215, assuming an exchange rate of 1 French franc in 1999 equals $0.20 in 2006.

38. Mycle Schneider, personal communication, July 22, 2013.

39. Aileen Mioko Smith, "Letters Sent to Countries Potentially on the Route of the MOX Fuel Shipment" (Green Action, Kyoto, March 5, 2013), www .greenaction-japan.org.

40. Feiveson et al., *Managing Spent Fuel from Power Reactors: Experience and Lessons from Around the World*.

41. The only repository under consideration containing oxygen is the proposed U.S. repository in Yucca Mountain, Nevada, which would be located above the ground water level.

42. *Nuclear Wastes: Technologies for Separations and Transmutation* (Washington, DC: National Academy Press, 1996), 3.

43. R. Bari et al., *Proliferation Risk Reduction Study of Alternative Spent Fuel Processing*, BNL-90264-2009-CP (Upton, NY: Brookhaven National Laboratory, July 2009).

44. Mycle Schneider and Yves Marignac, *Spent Nuclear Fuel Reprocessing in France* (Princeton, NJ: International Panel on Fissile Materials, April 2008), 30–40.

45. Vitrification has proven to be difficult in practice, leading to huge and dangerous accumulations of high-level liquid waste. In 2011, the UK Office of Nuclear Regulation, which has been the most open about this problem, proposed that the quantity of dissolved fission products from oxide fuel stored at the Sellafield reprocessing site in liquid form be limited to the equivalent of that in 2,000 tons of spent fuel, and that the total, including fission products from uranium metal "Magnox" fuel, be limited to the equivalent of 5,500 tons of spent fuel. *HAL [High Active Liquor] Stocks, Specification No. 793, Project Assessment Report* (UK Health and Safety Executive, Office of for Nuclear Regulation, 2011), 27. This is roughly equivalent to a limit of 140 megacuries (MCi) of cesium-137, about seventy times the amount released by the Chernobyl accident. (Assumes 4,000 MWt-days of fission per ton of uranium in Magnox fuel, 20,000 MWt-days per ton of uranium in Advanced Gas Reactor oxide fuel, 3.2 Ci of

cesium-137 produced per MWt-day, and, on average, ten years of decay since reactor discharge.)

46. The spent MOX would contain about 60 percent as much plutonium and americium as would have been put in the repository had there been no reprocessing and plutonium recycle.

47. Roald Wigeland, *Repository Benefit Analysis*, ANL-AFCI-089 (Argonne National Laboratory, 2003), 14.

48. *Trial Calculation of Nuclear Fuel Cycle Cost, Discussion Paper* (Japan Atomic Energy Commission, Subcommittee on Nuclear Power and Fuel Cycle Options, October 25, 2011); *Nuclear Decommissioning Authority: Managing Risk at Sellafield* (House of Commons Committee of Public Accounts, 2013), 3.

49. "Sellafield: Plant With an Image Problem," *BBC News*, July 23, 1998.

50. Assuming a zero discount rate, "Japan AEC Finds Direct Disposal Less Costly than Reprocessing," *Nuclear Intelligence Weekly*, November 28, 2011.

51. This estimates assumes 800 tons of spent fuel reprocessed annually for forty years, an average cumulative heat release of the spent fuel is 45 MWt-days per kilogram of heavy metal (kgHM), and a reactor heat-to-electric energy conversion ratio of 1/3.

52. *Global Fissile Material Report 2010*, chapters 2 and 3.

53. Fiona Harvey, "Sellafield MOX Nuclear Fuel Plant to Close," *The Guardian*, August 3, 2011; Steve Connor, "How a Money-Making Strategy From the 1960s Left Behind a Toxic Legacy," *The Independent*, April 11, 2011.

54. *Management of the UK's Plutonium Stocks: A Consultation Response on the Long-Term Management of UK-Owned Separated Civil Plutonium* (UK Department of Energy and Climate Change, December 2011).

7   Ending the Use of HEU as a Reactor Fuel

1. J. E. Matos, "Technical Challenges for Conversion of Civilian Research Reactors in Russia" (paper presented at the NAS-RAS Research Reactor Committee Briefing, National Academy of Sciences, Washington, DC, November 29, 2010).

2. *U.S. Nuclear Weapons Complex: Security at Risk* (Washington, DC: Project on Government Oversight, 2001).

3. Communiqué of the Washington Nuclear Security Summit, April 13, 2010, fpc.state.gov/documents/organization/140355.pdf.

4. Oleg Bukharin and William Potter, "Potatoes Were Guarded Better," *Bulletin of the Atomic Scientists*, June 1995.

5. Victor Gilinsky and Roger J. Mattson, "Revisiting the NUMEC Affair," *Bulletin of the Atomic Scientists* 66, no. 2 (April 2010).

6. In 1997, the U.S. Navy projected that the inventory of spent naval fuel would contain 65 metric tons of HEU in 2035. David Curtis, "Naval Nuclear Propulsion Program; Introduction to Spent Naval Fuel" (presented at the U.S. Nuclear

Waste Technical Review Board, Meeting of the Panel on the Repository, Augusta, GA, December 17, 1997), www.nwtrb.gov. Given that reprocessing of U.S. naval reactor fuel ended in 1992 and assuming that 40 percent of the uranium-235 in the 97.3 percent enriched fuel was fissioned, this would correspond to an average annual discharge of 2.5 tons. We assume that this rate is somewhat larger than the rate of fabrication of new cores because of the reduced number of nuclear submarines. See also *Global Fissile Material Report 2010: Balancing the Books, Production and Stocks* (Princeton, NJ: International Panel on Fissile Materials, 2010), 32, www.fissilematerials.org/library/gfmr10.pdf.

7. Gary Person, Dale Davis, and Russ Schmidt, "Progress Down-Blending Surplus Highly Enriched Uranium" (presented at the 50th Annual INMM Meeting, Institute of Nuclear Materials Management, Tucson, AZ, 2009).

8. For information on the global fleet of research reactors, see IAEA Research Reactor Database, nucleus.iaea.org/RRDB.

9. At the first Atoms for Peace conference held in Geneva in 1955, Alvin Weinberg, then Director of the U.S. Atomic Energy Commission's Oak Ridge National Laboratory, reported that he had "just received information from my country that sample UO_2-aluminum 20 per cent enriched fuel elements of the type which will be available to foreign countries have now been tested both in the LITR and in the MTR," that is, two research reactors operating at Oak Ridge National Laboratory. Session 9A, Volume II, August 12, 1955, p. 430, in *Proceedings of the International Conference on the Peaceful Uses of Atomic Energy Held in Geneva, August 8–20, 1955* (New York: United Nations, 1956).

10. *Highly Enriched Uranium: Striking a Balance. A Historical Report on the United States Highly Enriched Uranium Production, Acquisition, and Utilization Activities from 1945 through September 30, 1996* (Washington, DC: U.S. Department of Energy, December 2005), fig. 6-3, www.fissilematerials.org/library/doe01rev.pdf.

11. N. V. Arkhangelskiy, "Problems of the Research Reactors Conversion from HEU to LEU: History and Perspectives" (presented at the Russian-American Symposium on the Conversion of Research Reactors to LEU, Moscow, June 8, 2011).

12. Twenty-five non-weapon states with HEU-fueled research reactors are shown in *Global Fissile Material Report 2010*, chap. 11. All nine weapon states also had HEU-fueled research reactors by then. India, Israel, North Korea and Pakistan all were provided HEU-fueled reactors before they developed nuclear weapons: India (Apsara, 1956), Israel (IRR-1, 1960), North Korea (IRT-DPRK, 1965), and Pakistan (PARR-1, 1965). IAEA Research Reactor Database, nucleus.iaea.org/RRDB.

13. *Summary Volume, International Nuclear Fuel Cycle Evaluation (INFCE)* (Vienna: International Atomic Energy Agency, 1980), 40, 255–257.

14. For an early systematic discussion, see R. G. Muranaka, "Conversion of Research Reactors to Low-Enrichment Uranium Fuels," *IAEA Bulletin* 25, no. 1 (1984).

15. "Limiting the Use of Highly Enriched Uranium in Domestically Licensed Research and Test Reactors," *Federal Register* 51, no. 37 (U.S. Nuclear Regulatory Commission, February 25, 1986).

16. Matos, "Technical Challenges for Conversion of Civilian Research Reactors in Russia."

17. D. M. Wachs, *RERTR Fuel Development and Qualification Plan* (Idaho National Laboratory, INL/EXT-05-01017, Revision 3), January 2007.

18. "The Russian Reactor Conversion Program Might Begin to Work in 2013," *RIA Novosti*, October 1, 2012, http://ria.ru.

19. S. V. Kiriyenko and D. Poneman, "Joint Statement of the Co-Chairs of the Nuclear Energy and Nuclear Security Working Group of the Bilateral U.S.-Russia Presidential Commission," June 27, 2013.

20. A list of Russia's HEU-fueled research reactors may be found on the IPFM website at www.fissilematerials.org/facilities/research_reactors/russia.html.

21. Energy Policy Act of 1992, Sec. 903, "Restrictions on Nuclear Exports."

22. Arkhangelskiy, "Problems of the Research Reactors Conversion from HEU to LEU: History and Perspectives."

23. This is physically possible because the HEU density in the uranium containing "meat" inside the aluminum cladding of the pre-conversion fuel is almost always less than 1.7 g(U)/cc (grams of uranium per cubic centimeter) while the density of fuels under development range up to 15 g(U)/cc. Wachs, *RERTR Fuel Development and Qualification Plan*. The "meat" of first generation of higher density fuels has particles of various uranium compounds dispersed in an aluminum matrix. By 1988, densities of 4.8 g(U)/cc had been achieved with particles of U_3Si_2. Matos, "Technical Challenges for Conversion of Civilian Research Reactors in Russia."

24. Alexander Glaser and Uwe Filges, "Neutron-Use Optimization with Virtual Experiments to Facilitate Research-Reactor Conversion to Low-Enriched Fuel," *Science & Global Security* 20, nos. 2–3 (2012): 141–154.

25. N. E. Woolstenhulme et al., "Recent Accomplishments in the Irradiation Testing of Engineering-Scale Monolithic Fuel Specimens" (presented at the 34th International Meeting on Reduced Enrichment for Research and Test Reactors, Warsaw, Poland, 2012).

26. Thirty-eight reactors were converted prior to May 2004 and an additional twenty-seven by 2013. Armando Travelli, "Status and Progress of the RERTR Program in the Year 2004" (presented at the 26th International Meeting on Reduced Enrichment for Research and Test Reactors, Vienna, 2004). Jeff Chamberlin, U.S. Department of Energy, Global Threat Reduction Initiative, personal communication, April 11, 2013.

27. Ole Reistad and Styrkaar Hustveit, "HEU Fuel Cycle Inventories and Progress on Global Minimization," *Nonproliferation Review* 15, no. 2 (July 2008). Update through 2010 by Styrkaar Hustveit, personal communication, August 17, 2011. The list includes eighty-two steady-state reactors, eight pulsed reactors, and thirty-seven critical assemblies.

28. The Chinese built Miniature Neutron Source reactors (MNSR) and the Canadian SLOWPOKE reactors contain 0.9–1.0 kilograms of HEU in their cores.

29. Andrew Bieniawski, "An Overview of the NNSA Global Threat Reduction Programs Goals for HEU Minimization" (presented at the 31st International Meeting on Reduced Enrichment for Research and Test Reactors, Beijing, China, 2009).

30. *Non-HEU Production Technologies for Molybdenum-99 and Technetium-99m*, IAEA Nuclear Energy Series NF-T-5.4 (Vienna: International Atomic Energy Agency, 2013), 10–11.

31. *International Handbook of Evaluated Criticality Safety Benchmark Experiments* (Paris: OECD Nuclear Energy Agency, September 2012).

32. "Rolls-Royce Declines to Comment on Vulcan Future," *BBC*, November 3, 2011. The Vulcan submarine reactor site at Dounreay in Scotland was built in the 1950s and is to be shut down after 2015.

33. Harold F. McFarlane, "Is It Time to Consider Global Sharing of Integral Physics Data?" *Journal of Nuclear Science and Technology* 44, no. 3 (2007): 518–521.

34. The fifth U.S. HEU-fueled critical assembly is associated with the Advanced Test Reactor at the Idaho National Laboratory.

35. Most "sensitive" nuclear facilities in Russia (a list is not publicly available) are guarded by an Interior Ministry "intra-agency guard force" or interior troops. Pavel Podvig, personal communication, August 28, 2011.

36. The HEU-fueled Transient Reactor Test (TREAT) reactor at the Idaho National Laboratory, which has been used to test the behavior of nuclear fuels during a rapid buildup of power, has been on "cold standby" for more than a decade. *Ten-Year Site Plan, 2014–2023*, DOE/ID-11474 (Idaho National Laboratory, June 2012), www.inl.gov/publications/d/ten-year-site-plan.pdf. If TREAT is used again, it will require refurbishment and modernization. Its fuel is uranium oxide dispersed in graphite at a concentration of about 0.01 atom percent. This level of dilution does not make an attractive target for theft. J. H. Handwerk and R. C. Lied, *The Manufacture of the Graphite-Urania Fuel Matrix for TREAT*, ANL-5963 (Argonne National Laboratory, 1960).

37. "Alternative Source for Neutron Generation" (U.S. Department of Defense Small Business Innovation Research, 2012), www.dodsbir.net.

38. S. M. Myers, "Qualification Alternatives to the Sandia Pulsed Reactor (QASPR) Program Science in Sandia's Physical, Chemical, & Nano Sciences Center (PCNSC): Overview," *Research Briefs, Sandia National Laboratories*, 2007.

39. Edward J. Parma et al., *Operational Aspects of an Externally Driven Neutron Multiplier Assembly Concept Using a Z-Pinch 14-MeV Neutron Source (ZEDNA)* (Sandia National Laboratories, September 2007).

40. *INL's 52 Reactors*, Idaho National Laboratory, nuclear.inel.gov/52reactors .shtml.

41. Chunyan Ma and Frank von Hippel, "Ending the Production of Highly Enriched Uranium for Naval Reactors," *Nonproliferation Review* 8, no. 1 (Spring 2001): 86–101.

42. The United States only produced HEU with an enrichment greater than 96 percent uranium-235 at the Portsmouth gaseous diffusion plant. Production on a regular basis started in 1964 with an enrichment of over 97 percent and continued until 1992. The official history of U.S. HEU production enrichment states that "Until the mid-1960s, the plant produced HEU for nuclear weapons, the Naval Nuclear Propulsion Program, and other defense needs. With defense needs satisfied, the U.S. ceased the production of HEU for weapons in 1964. . . . After 1964, the U.S. continued to make HEU at Portsmouth for naval, space and research reactors." See p. 64 and table 5.4 in *Highly Enriched Uranium: Striking a Balance. A Historical Report on the United States Highly Enriched Uranium Production, Acquisition, and Utilization Activities from 1945 through September 30, 1996.*

43. After the collapse of the Soviet Union in 1991, Russia's submarines spent little time at sea and were not refueled for more than twenty years. See table 4.2 in *Global Fissile Material Report 2010.*

44. Ibid., 123. See also M. V. Ramana, "An Estimate of India's Uranium Enrichment Capacity," *Science & Global Security* 12, nos. 1–2 (2004): 115–124.

45. "France's Future SSNs: The Barracuda Class," *Defense Industry Daily*, December 21, 2011, www.defenseindustrydaily.com. France's *Georges Besse II* enrichment plant is licensed to produce LEU enriched up to 6 percent.

46. *Global Fissile Material Report 2010*, 101.

47. Leonam dos Santos Guimaraes, personal communication, July 2011.

48. In 2013, Russia had four nuclear-powered *Arktika*-class ocean icebreakers: the *Rossiya* (launched in 1985), the *Sovietsky Soyuz* (1990), the *Yamal* (1993) and the *50 Let Pobedy* (*50-Year Victory*, 2007), each equipped with twin 171 MWt reactors. Its three smaller ships—the river icebreakers *Taimyr* (1989) and *Vaigach* (1990) and the container ship *Sevmorput* (1988)—were equipped with single 135 MWt reactors. All were fueled with HEU. Anna Kireeva and Rashid Alimov, "Rosatom Takes over Russia's Nuclear Powered Icebreaker Fleet," *Bellona Foundation*, August 28, 2008, www.bellona.org.

49. *Global Fissile Material Report 2010*, 60–63.

50. Table 2 in Alexander Nikitin and Leonid Andreyev, *Floating Nuclear Power Plants* (Bellona Foundation, 2011).

51. "Small Nuclear Reactors for Power and Icebreaking," *World Nuclear News*, October 7, 2011.

52. Fiscal Year 1995 Defense Authorization Act, Public Law 103–337, Section 1042.

53. Director, Naval Nuclear Propulsion, *Report on Use of Low Enriched Uranium in Naval Nuclear Propulsion* (Washington, DC: U.S. Department of Energy, June 1995), 10, www.fissilematerials.org/library/onnp95.pdf.

54. The report explains that the ratio is not 4.7, as might be expected from the ratio of uranium-235 in 20 percent and 93.5 percent enriched uranium because of "fissioning of ^{238}U and its fissile transmutation products, ^{239}Pu and ^{241}Pu, and accounting for the lower fissions per unit volume in the fuel." Director, Naval

Nuclear Propulsion, *Report on Use of Low Enriched Uranium in Naval Nuclear Propulsion* (Washington, DC: U.S. Department of Energy, June 1995), 10, www.fissilematerials.org/library/onnp95.pdf.

55. For an attack submarine, the additional cost would be $400 million, of which $250 million would be for "government-furnished reactor core and reactor heavy equipment." The remainder would be for increasing the size of the submarine. The larger ballistic-missile submarines and aircraft carriers would not have to be increased in size to accommodate larger cores. The additional cost for the ballistic missile submarine would be $325 million "for the reactor equipment and core" and, for an aircraft carrier, "$1.23 billion more in government-furnished reactor core and reactor heavy equipment." It was assumed that, on average, 1.5 attack submarines, one third of a ballistic-missile submarine and one quarter of an aircraft carrier would be built per year (ibid., 24). The cost of the fuel fabrication is given for the case in which the reactor cores are kept the same size but replaced three times in the case of the submarines and twice in the case of the aircraft carriers: $85 million/core for an attack submarine core, $115 million for ballistic missile submarines, and $590 million for two aircraft carrier cores. Assuming 0.7 tons of HEU per attack submarine, 1 ton per ballistic missile submarine, and 5 tons per aircraft carrier, this cost would amount to about $100,000 per kilogram of uranium. This is about the same as the cost of the High Flux Isotope Reactor. J. D. Sease et al., *Conceptual Process for the Manufacture of Low-Enriched Uranium/Molybdenum Fuel for the High Flux Isotope Reactor*, ORNL/TM-2007/39 (Oak Ridge National Laboratory, 2007). However, the design of naval reactor cores is much less complex.

56. *Report on Use of Low Enriched Uranium in Naval Nuclear Propulsion*, 1.

57. *FY 2013 Congressional Budget Request* (Washington, DC: U.S. Department of Energy, 2012), 484.

58. "France's Future SSNs: The Barracuda Class."

59. Y. Girard (presented at the Conference on the Implication of the Acquisition of Nuclear-Powered Submarines [SSN] by Non-Nuclear Weapons States, Massachusetts Institute of Technology, Cambridge, MA, March 27, 1989).

60. Office of Naval Reactors, *Report on Low Enriched Uranium for Naval Reactor Cores* (Washington, DC: U.S. Department of Energy, January 2014), www.fissilematerials.org/library/doe14.pdf.

8   Ending Production of Fissile Materials for Weapons

1. *General and Complete Disarmament*, Resolution A/RES/48/75 (New York: United Nations General Assembly, December 16, 1993), www.un.org/documents/ga/res/48/a48r075.htm.

2. International Panel on Fissile Materials, "A Fissile Material (Cut-Off) Treaty— A Treaty Banning the Production of Fissile Materials for Nuclear Weapons or Other Nuclear Explosive Devices, with article-by-article explanations," September 2, 2009, www.fissilematerials.org/library/G1060052.pdf.

3. *Decision for the Establishment of a Programme of Work for the 2009 Session*, Resolution CD/1864 (Geneva: Conference on Disarmament, May 29, 2009).

4. Zia Mian and A. H. Nayyar, "Playing the Nuclear Game: Pakistan and the Fissile Material Cutoff Treaty," *Arms Control Today*, April 2010. Pakistan's ambassador to the CD, Zamir Akram, argued in 2011, however, that if Pakistan were given the same waiver from the Nuclear Suppliers Group (NSG) ban on nuclear trade with countries not party to the Non-Proliferation Treaty that had been granted to India in 2008, Pakistan would allow FMCT negotiations to start. "The South Asian Nuclear Balance: An Interview With Pakistani Ambassador to the CD Zamir Akram," *Arms Control Today*, December 2011. There was no indication that the five NPT weapon states would support Pakistan's demand for an NSG waiver.

5. Aluf Benn, "The Struggle to Keep Nuclear Capabilities Secret," *Haaretz*, September 14, 1999. See also "Israel" in *Banning the Production of Fissile Materials for Nuclear Weapons: Country Perspectives on the Challenges to a Fissile Material (Cutoff) Treaty* (International Panel on Fissile Materials, September 2008), www.fissilematerials.org/library/gfmr08cv.pdf.

6. *United States Memorandum Submitted to the First Committee of the General Assembly*, January 12, 1957, www.gwu.edu/~nsarchiv/nukevault/ebb321/3A .pdf.

7. China reportedly wishes to maintain the option of increasing the size of its nuclear arsenal in case the United States builds up its long-range conventional strike capabilities and ballistic-missile defenses, bringing China's nuclear deterrent into question. Li Bin, "China," in *Banning the Production of Fissile Materials for Nuclear Weapons: Country Perspectives on the Challenges to a Fissile Material (Cutoff) Treaty*.

8. As of the end of 2012, the Czech Republic, France, Japan, South Korea, Norway, Pakistan, Taiwan, the United Kingdom, the United States, and the European Commission submit information about separated neptunium and americium to the International Atomic Energy Agency. *Safeguards Statement for 2012* (Vienna: International Atomic Energy Agency, 2013), www.iaea.org/safeguards/ es/es2012.html.

9. Paragraph 15.10 in *2000 Review Conference of the Parties to the Treaty on the Non-Proliferation of Nuclear Weapons, Final Document, Parts I and II*, NPT/ CONF.2000/28 (New York: United Nations, 2000), www.un.org/disarmament/ WMD/Nuclear/2000-NPT/2000NPTDocs.shtml.

10. The 2010 NPT Review Conference only reported agreement that "the nuclear-weapon States are encouraged . . . to declare . . . to the International Atomic Energy Agency (IAEA) all fissile material designated by each of them as no longer required for military purposes and to place such material as soon as practicable under IAEA or other relevant, international verification," *2010 Review Conference of the Parties to the Treaty on the Non-Proliferation of Nuclear Weapons, Final Document, Volume 1* (New York: United Nations, 2010), Part I, Action 16, www.un.org/en/conf/npt/2010.

11. Article VII.3 of the agreement states that "each Party, in cooperation with the other Party, shall begin consultations with the International Atomic Energy

Agency (IAEA) at an early date and undertake all other necessary steps to conclude appropriate agreements with the IAEA to allow it to implement verification measures with respect to each Party's disposition program." "2000 Plutonium Management and Disposition Agreement as Amended by the 2010 Protocol," April 2010, www.fissilematerials.org/library/PMDA2010.pdf. The original 2000 agreement had a similar but more detailed requirement.

12. The chairman of India's Department of Atomic Energy declared that "both from the point of view of maintaining long term energy security and for maintaining the minimum credible deterrent the Fast Breeder Programme just cannot be put on the civilian list." "On the Fast Breeder Programme, Begin a Civil Debate," *The Indian Express*, February 10, 2006. This suggests to some that India was leaving open the option of using the breeder to produce weapon grade plutonium. India could produce about 140 kilograms of weapon-grade plutonium per year by fueling the core of its 500 MWe Prototype Fast Reactor with reactor-grade plutonium and producing weapon-grade plutonium in the uranium blanket around the core. Alexander Glaser and M. V. Ramana, "Weapon-Grade Plutonium Production Potential in the Indian Prototype Fast Breeder Reactor," *Science & Global Security* 15, no. 2 (2007): 85–105.

13. Gary Person, Dale Davis, and Russ Schmidt, "Progress Down-Blending Surplus Highly Enriched Uranium" (presented at the 53rd Annual INMM Meeting, Institute of Nuclear Materials Management, Orlando, FL, 2012).

14. Statement by Ambassador Zamir Akram, Permanent Representative of Pakistan at the Conference on Disarmament, Geneva, February 18, 2010.

15. David Albright and Robert Avagyan, *Construction Progressing Rapidly on the Fourth Heavy Water Reactor at the Khushab Nuclear Site* (Washington, DC: Institute for Science and International Security, May 21, 2012); David Albright and Paul Brannan, *Pakistan Expanding Plutonium Separation Facility Near Rawalpindi* (Institute for Science and International Security, May 19, 2009).

16. *Treaty on the Non-Proliferation of Nuclear Weapons*, Article III.

17. Pakistan, however, is considering the need for a nuclear submarine program to respond to India. Tauquir H. Naqvi, "Indian Nuclear Submarine Programme," *The Nation*, May 13, 2012. Pakistan may not need to build up a stockpile of HEU for naval propulsion, however, if it follows China's example and uses LEU fuel for its nuclear submarine reactors.

18. A. Glaser and S. Bürger, "Verification of a Fissile Material Cutoff Treaty: The Case of Enrichment Facilities and the Role of Ultra-Trace Level Isotope Ratio Analysis," *Journal of Radioanalytical and Nuclear Chemistry* 280, no. 1 (April 2009): 85–90.

19. Shirley Johnson, *Safeguards at Reprocessing Plants Under a Fissile Material (Cutoff) Treaty*, Research Report 6 (Princeton, NJ: International Panel on Fissile Materials, February 2009), www.fissilematerials.org/library/rr06.pdf.

20. Ibid.

21. *High Levels of Radioactive Carbon Contamination Around La Hague* (Greenpeace, November 12, 1998); C. Fréchou and D. Calmet, "^{129}I in the Environment of the La Hague Nuclear Fuel Reprocessing Plant—from Sea to Land,"

Journal of Environmental Radioactivity 70, nos. 1–2 (2003): 43–59. According to Areva's gaseous release annual statement made available on the company's website, in 2012, 0.14 Curies (Ci) of radioiodine, 440 Ci of carbon-14, and 5.8 million Ci of noble gases were released at La Hague, www.areva.com.

22. R. Scott Kemp, "A Performance Estimate for the Detection of Undeclared Nuclear-fuel Reprocessing by Atmospheric 85Kr," *Journal of Environmental Radioactivity* 99, no. 8 (August 2008): 1341–1348. A rate of production of weapon-grade plutonium of 20 kilograms per year would correspond to an average release rate of about 800 gigabecquerels (GBq) of krypton-85 per day.

23. One significant quantity is defined by the IAEA as "the approximate amount of nuclear material for which the possibility of manufacturing a nuclear explosive device cannot be excluded." Table II in *IAEA Safeguards Glossary, 2001 Edition* (Vienna: International Atomic Energy Agency, 2003). This number is appropriate for a first-generation (Nagasaki-type) implosion bomb including production losses.

24. For example, the IAEA has conducted challenge inspections at a number of Iran's military facilities. *Communication Dated 12 September 2005 from the Permanent Mission of the Islamic Republic of Iran to the Agency*, INFCIRC/657 (Vienna: International Atomic Energy Agency, September 15, 2005), 11–12.

25. Part X. "Challenge Inspections Pursuant to Article IX," Paragraphs 46–48, in *Convention on the Prohibition of the Development, Production, Stockpiling and Use of Chemical Weapons and on Their Destruction (Chemical Weapons Convention)* (The Hague, the Netherlands: Organization for the Prohibition of Chemical Weapons, 2005).

26. Information about the Automated Mass Spectral Deconvolution and Identification System (AMDIS) developed by the U.S. government and examples of gas-chromatographic mass-spectrometer data can be found at chemdata.nist.gov.

27. Report by the Director General, *Strengthening the Effectiveness and Improving the Efficiency of the Safeguards System and of the Model Additional Protocol*, GC(54)/11 (Vienna: International Atomic Energy Agency, July 27, 2010), para. 27.

28. *Global Fissile Material Report 2010: Balancing the Books, Production and Stocks* (Princeton, NJ: International Panel on Fissile Materials, 2010), www.fissilematerials.org/library/gfmr10.pdf. The naval reactors in Chinese and French submarines and Brazil's prototype naval reactor are believed to use LEU fuel.

29. The only information that has been declassified is that zirconium alloy is used in U.S. naval reactor cores. *Restricted Data Declassification Decisions 1946 to the Present (RDD-8)* (U.S. Department of Energy, January 1, 2002), www.osti.gov.

30. Personal communication, Leonam dos Santos Guimaraes, July 2011.

31. Article 14, "Non-application of safeguards to nuclear material to be used in non-peaceful activities," in *The Structure and Content of Agreements between the Agency and States Required in Connection with the Treaty on the Non-Proliferation of Nuclear Weapons*, INFCIRC/153 (Corrected) (Vienna: International Atomic Energy Agency, June 1972).

32. According to the official history of the IAEA, this loophole was introduced because of Italy's interest at the time of the NPT negotiations in building a nuclear-powered supply ship and the Netherlands in building a nuclear submarine. David Fischer, *History of the International Atomic Energy Agency: The First Forty Years* (Vienna: IAEA, 1997), 272–273.

33. *FY 2013 Congressional Budget Request* (Washington, DC: U.S. Department of Energy, 2012), Volume 1, 484.

34. *Highly Enriched Uranium: Striking a Balance. A Historical Report on the United States Highly Enriched Uranium Production, Acquisition, and Utilization Activities from 1945 through September 30, 1996* (Washington, DC: U.S. Department of Energy, December 2005), 39, www.fissilematerials.org/library/doe01rev.pdf.

9   Disposal of Fissile Materials

1. As of the end of 2012, the United Kingdom held about 96 tons of domestic and 24 tons of foreign owned separated civilian plutonium. The UK stock will increase to about 100 tons if outstanding reprocessing contracts are fulfilled before its reprocessing plants are shut down.

2. The IAEA has exempted plutonium containing more than 80 percent plutonium-238 from safeguards because it generates a large amount of radioactive decay heat due to its relatively short half-life (88 years), making its use in weapons difficult. Because of its heat generation, plutonium-238 is used to power thermoelectric generators used in deep space probes.

3. *Management and Disposition of Excess Weapons Plutonium* (Washington, DC: U.S. National Academy of Sciences, 1994), 34.

4. This would require mixing the plutonium with some material other than uranium-238, since the neutron irradiation of uranium-238 generates new plutonium.

5. Steve Fetter and Frank von Hippel, "The Hazard from Plutonium Dispersal by Nuclear-warhead Accidents," *Science & Global Security* 2 (1990): 21–41.

6. Risk projections suggest that by 2065, the Chernobyl disaster may produce about 16,000 cases of thyroid cancer and 25,000 cases of other cancers across Europe, with about half of these expected in the most contaminated areas of Belarus, Russia, and Ukraine, but buried in a background of many more cancer cases expected from other causes. Elisabeth Cardis et al., "Estimates of the Cancer Burden in Europe from Radioactive Fallout from the Chernobyl Accident," *International Journal of Cancer* 119, no. 6 (2006): 1224–1235. Psychological and other social effects on the populations from the more contaminated areas are described in "Exposures and Effects of the Chernobyl Accident," Annex J in *Sources and Effects of Ionizing Radiation* (New York: United Nations Scientific Committee on the Effects of Atomic Radiation [UNSCEAR], Report to the General Assembly, 2000), 513–514.

7. Adolf von Baeckmann, Garry Dillon, and Demetrius Perricos, "Nuclear Verification in South Africa," *IAEA Bulletin* 37, no. 1 (1995): 42–48.

8. For estimates of HEU and plutonium consumed by various weapon states in nuclear weapons tests, see the respective country chapters in *Global Fissile Material Report 2010: Balancing the Books, Production and Stocks* (Princeton, NJ: International Panel on Fissile Materials, 2010), www.fissilematerials.org/library/gfmr10.pdf.

9. Thomas L. Neff, "A Grand Uranium Bargain," *New York Times*, October 24, 1991.

10. James P. Timbie, "Energy from Bombs: Problems and Solutions in the Implementation of a High-Priority Nonproliferation Project," *Science & Global Security* 12, no. 3 (2004): 165–189.

11. The original framework "Agreement between the Government of the United States of America and the government of the Russian Federation Concerning the Disposition of Highly Enriched Uranium Extracted from Nuclear Weapons" was signed in Washington on February 18, 1993, www.fissilematerials.org/library/heu93.pdf.

12. Uranium-234 is a decay product of uranium-238. It has a half-life of 245,000 years, 1/20,000 that of uranium-238, and is therefore 20,000 times more radioactive per gram. Because it is lighter than U-235, an even larger fraction of U-234 is extracted from natural uranium by gaseous diffusion and gas centrifuge enrichment processes. See *Options for Expanding Conversion of Russian Highly Enriched Uranium* (Nuclear Threat Initiative, December 2010), 7.

13. Pavel Podvig, "The Fallacy of the Megatons to Megawatts Program," *Bulletin of the Atomic Scientists*, July 23, 2008, thebulletin.org/fallacy-megatons-megawatts-program; Pavel Podvig, *Consolidating Fissile Materials in Russia's Nuclear Complex*, IPFM Research Report 7 (Princeton, NJ: International Panel on Fissile Materials, May 2009), www.fissilematerials.org/library/rr07.pdf.

14. "U.S. Uranium Down-Blending Activities: Fact Sheet" (U.S. Department of Energy, National Nuclear Security Administration, March 23, 2012).

15. Nuclear Threat Initiative, *Options for Expanding Conversion of Russian Highly Enriched Uranium*, December 2010.

16. One of the reactor-produced uranium isotopes is uranium-232, which has a half-life of 69 years. One of its decay products, thallium-208, produces a high-energy gamma ray when it decays. A second reactor-produced uranium isotope, uranium-236 (half-life 23 million years) has a large cross-section for neutron capture that reduces the reactivity of fuel and must be offset by raising the uranium-235 enrichment.

17. In 1995, the United States declared 174 tons of mostly non-weapon-grade HEU excess for all military (weapon and naval) purposes. Most of it has been or is to be disposed of by blend-down to LEU. About 21 tons in spent fuel are to be disposed as waste. In 2005, the U.S. Secretary of Energy announced an additional 200 tons of HEU would be removed from the weapons arsenal, and

that approximately 20 tons of this material, later increased to about 28 tons, would be down-blended to LEU and that another 20 tons would be reserved for research reactor and space reactor fuel. As of 2012, 152 tons of weapon-grade uranium were reserved for naval propulsion reactor fuel. Gary Person, Dale Davis, and Russ Schmidt, "Progress Down-Blending Surplus Highly Enriched Uranium" (presented at the 53rd Annual INMM Meeting, Institute of Nuclear Materials Management, Orlando, FL, 2012). The final down-blend to the exact enrichment required is done at the fuel fabrication plants.

18. Babcock and Wilcox operates the Y-12 facility for the U.S. government and owns facilities at Lynchburg, Virginia, and Erwin, Tennessee, where naval and research reactor fuel is produced and excess HEU is blended down.

19. *FY 2015 Congressional Budget Request* (Washington, DC: U.S. Department of Energy, 2014), Vol. 1, 539.

20. Article I, "Agreement between the Government of the United States of America and the Government of the Russian Federation Concerning the Management and Disposition of Plutonium Designated as No Longer Required for Defense Purposes and Related Cooperation," 2000, www.fissilematerials.org/library/gov00.pdf.

21. *Communication Received from the United States of America Concerning Its Policies Regarding the Management of Plutonium*, INFCIRC/549/Add.6/14 (Vienna: International Atomic Energy Agency, October 19, 2012).

22. *Plutonium: The First 50 Years. United States Plutonium Production, Acquisition and Utilization from 1944 through 1994*, DOE/DP-0137 (Washington, DC: U.S. Department of Energy, February 1996), www.fissilematerials.org/library/doe96.pdf.

23. *Management and Disposition of Excess Weapons Plutonium*; *Management and Disposition of Excess Weapons: Reactor-Related Options* (Washington, DC: U.S. National Academy of Sciences, 1994 and 1995). These reports built on an earlier study by Frans Berkhout et al., "Disposition of Separated Plutonium," *Science & Global Security* 3, nos. 3–4 (1993): 161–213.

24. Weapon-grade plutonium is more than 90 percent plutonium-239, which replaces the uranium-235 in low-enriched uranium on an approximately 1.3 atom for one atom basis. *Plutonium Fuel: An Assessment. Report by an Expert Group* (Paris: OECD Nuclear Energy Agency, 1989), Tables 9 and 12. Civilian plutonium, however, contains about 35 percent plutonium isotopes that cannot sustain a slow-neutron chain reaction. Its percentage in MOX fuel must therefore be increased correspondingly to achieve the same fuel value.

25. "2000 Plutonium Management and Disposition Agreement as Amended by the 2010 Protocol," April 2010, www.fissilematerials.org/library/PMDA2010.pdf. The Protocol was signed by Secretary Clinton and Foreign Minister Lavrov on April 13, 2010. In the Protocol, the United States also committed to dispose of all 34 tons of its plutonium covered by the agreement in MOX.

26. Article VII.3, "Agreement between the Government of the United States of America and the Government of the Russian Federation Concerning the

Management and Disposition of Plutonium Designated as No Longer Required for Defense Purposes and Related Cooperation."

27. In 2001, a joint U.S.-Russian study envisioned that 14.5 tons of Russia's excess plutonium would be used in the BN-600 fast-neutron reactor and the rest in LWRs. *Cost Analysis and Economics in Plutonium Disposition: Cost Estimates for the Disposition of Weapon-Grade Plutonium Withdrawn from Military Programs* (Washington, DC: Joint U.S.-Russian Working Group on Cost Analysis and Economics in Plutonium Disposition, 2001).

28. The first U.S.-Russian estimate of the cost of the Russian plutonium disposition program via MOX was $0.8 billion upfront cost, including $0.3 billion for the MOX plant plus $1.1 billion for operating costs. *Preliminary Cost Assessment for the Disposition of Weapon-Grade Plutonium Withdrawn From Russia's Nuclear Military Programs* (Washington, DC: Joint U.S.-Russian Working Group on Cost Analysis and Economics in Plutonium Disposition, April 2000). In the April 2003 report of the Joint Working Group, the cost of the MOX plant had increased to $0.64 billion. It was explained that "the current assumption is that all major plant items and equipment would be procured from the same suppliers as, and at the same prices as, the equivalent items and equipment in the counterpart U.S. plant. *Scenarios and Costs in the Disposition of Weapon-Grade Plutonium Withdrawn from Russia's Nuclear Military Programs* (Washington, DC: Joint U.S.-Russian Working Group on Cost Analysis and Economics in Plutonium Disposition, 2003), 25.

29. The numerical suffix denotes the gross electrical generating capacity of a Russian power reactor, which is typically 7–10 percent more than its net generating capacity.

30. *Report to Congress: Disposition of Surplus Defense Plutonium at Savannah River Site* (Washington, DC: U.S. National Nuclear Security Administration, Office of Fissile Material Disposition, February 15, 2002).

31. *FY 2014 Congressional Budget Request* (Washington, DC: U.S. Department of Energy, 2013), Vol. 1. $7.7 billion construction cost for the Mixed Oxide Fuel Fabrication Facility (DN-119); $8.2 billion for operations and security costs over fifteen years (DN-147); $0.4 billion for the associated Waste Solidification Building and $1.9 billion for its operation over twenty years (DN-148).

32. Ibid., DN-119; Tom Clements, Edwin Lyman, and Frank von Hippel, "The Future of Plutonium Disposition," *Arms Control Today*, July/August 2013, 8–15.

33. Steve Connor, "How a Money-Making Strategy From the 1960s Left Behind a Toxic Legacy," *The Independent*, April 11, 2011.

34. Phillip Chaffee, "Sellafield Mox Plant to Close," *Nuclear Intelligence Weekly*, August 8, 2011, 5.

35. *Management of the UK's Plutonium Stocks: A Consultation Response on the Long-Term Management of UK-Owned Separated Civil Plutonium* (UK Department of Energy and Climate Change, December 1, 2011), 15. The official discount rate is 3.5 percent, and it is assumed that the MOX plant would

operate from 2025 to 2053. *Plutonium: Credible Options Analysis* (UK Nuclear Decommissioning Authority, 2010), 35.

36. *Management of the UK's Plutonium Stocks: A Consultation Response on the Long-Term Management of UK-Owned Separated Civil Plutonium*, 5.

37. Assuming that MOX constituted one-third of 20 tons of fuel loaded per year. British Energy has shown no interest in licensing its aging Advanced Gas-Cooled Reactors AGR reactor for MOX. *Plutonium: Credible Options Analysis*, 16.

38. J. W. Hobbs, "A Programme to Immobilise Plutonium Residues at Sellafield" (presented at the 53rd Annual INMM Meeting, Institute of Nuclear Materials Management, Orlando, FL, 2012).

39. Frank von Hippel et al., "Nuclear Proliferation: Time to Bury Plutonium," *Nature* 485, no. 7397 (2012): 167–168.

40. See, for example, W. J. Weber, R. C. Ewing, and W. Lutze, "Performance Assessment of Zircon as a Waste Form for Excess Weapons Plutonium under Deep Borehole Burial Conditions," *MRS Online Proceedings Library* 412 (1995); Neil Chapman and Fergus Gibb, "A Truly Final Waste Management Solution: Is Very Deep Borehole Disposal A Realistic Option For High-Level Waste Or Fissile Materials?" *Radwaste Solutions* 10, no. 4 (August 2003); Patrick Brady et al., *Deep Borehole Disposal of Nuclear Waste: Final Report*, SAND2012-7789 (Sandia National Laboratories, 2012); P. V. Brady and M. J. Driscoll, "Deep Borehole Disposal of Nuclear Waste," *Radwaste Solutions* 17, no. 5 (October 2010); *Project on Alternative Systems Study (PASS): Final Report*, TR 93-04 (Stockholm: SKB, October 1992); *Choice of Method—Evaluation of Strategies and Systems for Disposal of Spent Nuclear Fuel*, P-10-47 (Stockholm: SKB, October 2010); *A Review of the Deep Borehole Disposal Concept for Radioactive Waste*, N/108 (Oxfordshire: Nirex, June 2004).

41. Rod Ewing, "Geological Disposal" in Harold A. Feiveson et al., *Managing Spent Fuel from Power Reactors: Experience and Lessons from Around the World* (Princeton, NJ: International Panel on Fissile Materials, September 2011).

42. *A Review of the Deep Borehole Disposal Concept for Radioactive Waste*, 66.

43. *The United States Plutonium Balance, 1944–2009* (Washington, DC: U.S. Department of Energy, June 2012).

44. *Draft Surplus Plutonium Supplemental Environmental Impact Statement*, DOE/EIS00283-S2 (Washington, DC: U.S. Department of Energy, 2012), S–23, S–24, S–31, S–33.

45. David Sanger, "Obama to Renew Drive for Cuts in Nuclear Arms," *New York Times*, February 10, 2013.

46. Masafumi Takubo and Frank von Hippel, *Ending Reprocessing in Japan: An Alternative Approach to Managing Japan's Spent Fuel and Separated Plutonium* (Princeton, NJ: International Panel on Fissile Materials, 2013), www.fissilematerials.org/library/rr12.pdf.

10   Conclusion: Unmaking the Bomb

1. James Franck et al., *Report of the Committee on Political and Social Problems Manhattan Project* (The Franck Report) (University of Chicago, June 11, 1945), www.fissilematerials.org/library/fra45.pdf.

2. Ibid.

3. Ibid.

4. C. I. Barnard et al., *A Report on the International Control of Atomic Energy* (Washington, DC, 1946), www.fissilematerials.org/library/ach46.pdf.

5. Richard Rhodes, *Arsenals of Folly: Nuclear Weapons in the Cold War* (New York: Alfred A. Knopf, 2007).

6. For more detail, see *Global Fissile Material Report 2013: Increasing Transparency of Nuclear Warhead and Fissile Material Stocks as a Step Toward Disarmament* (Princeton, NJ: International Panel on Fissile Materials, November 2013), www.fissilematerials.org/library/gfmr13.pdf.

7. Action 3, *2010 Review Conference of the Parties to the Treaty on the Non-Proliferation of Nuclear Weapons, Final Document, Volume 1* (New York: United Nations, 2010), www.un.org/en/conf/npt/2010.

8. "Remarks by President Obama at Hankuk University" (Seoul, Republic of Korea, March 26, 2012).

9. David Sanger, "Obama to Renew Drive for Cuts in Nuclear Arms," *New York Times*, February 10, 2013.

10. Chapter 11 in Harold A. Feiveson, ed., *The Nuclear Turning Point: A Blueprint for Deep Cuts and De-alerting of Nuclear Weapons* (Washington, DC: Brookings Institution, 1999).

11. This approach was originally suggested by Steve Fetter, "Nuclear Archaeology: Verifying Declarations of Fissile-Material Production," *Science & Global Security* 3, nos. 3–4 (1993). For a discussion and a review of several case studies involving graphite-moderated plutonium production reactors, see T. W. Wood et al., "Establishing Confident Accounting for Russian Weapons Plutonium," *Nonproliferation Review* 9, no. 2 (Summer 2002): 126–137. Equivalent methods have been proposed for other types of reactors, especially for heavy-water-moderated reactors. Alex Gasner and Alexander Glaser, "Nuclear Archaeology for Heavy-Water-Moderated Plutonium Production Reactors," *Science & Global Security* 19, no. 3 (2011): 223–233.

12. For more details, see the biography by William Lanouette, *Genius in the Shadows: A Biography of Leo Szilard: The Man behind the Bomb* (New York: C. Scribner's Sons, 1992), especially chapter 8.

13. Szailard's proposal came in a March 1945 memo, "Atomic Bombs and the Postwar Position of the United States," meant for transmission to President Roosevelt. The memo is reproduced in Spencer R. Weart and Gertrud Weiss Szilard, *Leo Szilard: His Version of the Facts: Selected Recollections and Correspondence*

(Cambridge, MA: MIT Press, 1978), 196–204. President Roosevelt died before the memo could be delivered.

14. In 1995, Hans Bethe, the head of the theoretical division at Los Alamos during the Manhattan Project, issued an appeal to "all scientists in all countries to cease and desist from work creating, developing, improving and manufacturing further nuclear weapons." *Hans A. Bethe's Letter to the Science Community,* July 23, 1995, www.pugwash.org/about/bethe.htm.

15. Joseph Rotblat, "Nobel Lecture: Remember Your Humanity" (Oslo, Norway, December 10, 1995), www.nobelprize.org/nobel_prizes/peace/laureates/1995/rotblat-lecture.html. See also Joseph Rotblat, "Societal Verification," in Joseph Rotblat et al., *A Nuclear-Weapon-Free World: Desirable? Feasible?* (Boulder: Westview Press, 1993), 103–118.

16. For earlier suggestions in this direction, see Grenville Clark and Louis B. Sohn, *World Peace through World Law,* 2nd edition (Somerset, NJ: Transaction Publishers, 1960), 267. Also Lewis Bohn, "Non-Physical Inspection Techniques," in Donald G. Brennan, *Arms Control, Disarmament and National Security* (New York: G. Braziller, 1961).

17. Jonathan Schell, *The Abolition* (New York: Knopf, 1984).

18. Niels Bohr, "Memorandum to President Roosevelt," July 3, 1944, in Niels Bohr, ed., *Collected Works, Volume 11: The Political Arena (1934–1961)* (Amsterdam: Elsevier, 2005), 101–108.

19. Lawrence S. Wittner, *Confronting the Bomb: A Short History of the World Nuclear Disarmament Movement* (Stanford: Stanford University Press, 2009).

Glossary

Additional Protocol. The voluntary agreement by a state to accept more stringent International Atomic Energy Agency (IAEA) safeguards than those originally established to verify compliance with the Non-Proliferation Treaty. Devised in the 1990s following the discovery of Iraq's clandestine uranium enrichment programs, it broadens the information on nuclear activities a state declares to the IAEA and provides additional rights for IAEA inspectors to verify this declaration, including environmental sampling to check for possible undeclared nuclear activities in a country.

Americium-241 (Am-241). A fissile isotope with a half-life of 433 years produced from decay of plutonium-241. There is no public information that americium has ever been used in a deployed nuclear weapon but it has been used in weapons research and development, including nuclear weapon tests. It is considered an "alternative nuclear material" by the IAEA.

Boosting. A mechanism to increase the amount of fissile material consumed in a fission weapon explosion thus increasing the yield of the explosion. Boosting relies on the heat of the fission explosion to induce fusion of the heavy isotopes of hydrogen, deuterium and tritium, to produce additional neutrons that cause additional fissions.

Breeder reactor. A nuclear reactor designed to consume less chain-reacting fissile material in its core than it produces in a surrounding blanket of "fertile" uranium-238 (U-238) or thorium. Most research and development has been focused on fast-neutron plutonium breeder reactors cooled with liquid sodium. Sodium burns if exposed to air or water. This makes refueling and maintenance complex. Despite many attempts, breeder reactors have not been successfully commercialized.

Burnup. A measure of the fission energy generated by a mass of nuclear fuel in a reactor, usually given at the time of discharge from the reactor, measured in units of thermal megawatt-days per kilogram or thousand thermal megawatt-days per metric ton.

Calutron. A uranium isotope enrichment technology that relies on a magnetic field to separate a beam of uranium ions, with lighter uranium-235 ions being bent more than heavier uranium-238 ions.

CANDU reactor. Canadian designed heavy-water moderated and natural-uranium fueled power reactor.

Cascade. The arrangement of isotope separation elements (e.g., centrifuges) in a uranium enrichment facility. The cascade is organized as a series of "stages" in each of which the isotopic separation elements operate in parallel. The stages are connected so that material enriched in one stage is passed to the next for further enrichment of the uranium in the isotope uranium-235. If the feed into the cascade is natural uranium, the final output streams are enriched and depleted uranium.

Centrifuge. A rapidly rotating cylinder used for the enrichment of uranium in which the heavier isotope (uranium-238) in uranium hexafluoride gas is forced to higher concentrations near the cylinder's walls, while the lighter isotope (uranium-235) concentrates closer to the center of the cylinder.

Chain reaction. A continuing process of nuclear fissioning in which neutrons released from one fission trigger other nuclear fissions. In a nuclear weapon, an extremely rapid, multiplying chain reaction causes an explosive release of energy. In a reactor operating at constant power, the chain reaction is controlled so that each fission causes, on average, exactly one follow-on fission.

Critical mass. The minimum amount of a fissile material required to sustain a chain reaction. The exact mass of material needed to sustain a chain reaction varies according to its geometry, the mixture of fissile isotopes and other elements it contains, its density, and the neutron-reflecting properties and thickness of the surrounding materials.

Depleted uranium. Uranium having a smaller percentage of uranium-235 than the 0.7 percent found in natural uranium. It is the waste product of the uranium enrichment process.

Down-blending. The dilution of highly enriched uranium with depleted, natural, or slightly enriched uranium (known as blend-stock) to produce low-enriched uranium that can be used to fuel light water reactors. This has been the method for disposing of stocks of highly enriched uranium from excess Cold War nuclear weapons.

Enrichment. The process of increasing the fraction of one isotope of a given element; in the case of uranium, increasing the concentration of chain-reacting uranium-235 relative to the more abundant isotope uranium-238.

Environmental sampling. The set of techniques used by the IAEA to take swipes of surfaces within and around nuclear facilities and potentially to collect for analysis samples of air, water, soil, and vegetation in states that have signed the Additional Protocol of the NPT to detect indications of the undeclared production of fissile materials, such as the presence of plutonium, fission products, or highly enriched uranium.

Fast-neutron reactor (fast reactor). A type of nuclear reactor in which the chain reaction is sustained by fission neutrons whose kinetic energy and speed have been minimally reduced ("moderated"). It typically requires fuel that has a 20–30 percent concentration of plutonium or uranium-235 in uranium—much higher than in the fuel of reactors such as light water reactors in which the neutron energy is "moderated" to the level of the thermal motions of molecules in the reactor coolant. When the core of a fast reactor is surrounded by a blanket of uranium or thorium, it can produce more fissile material than it consumes and is known as a breeder reactor.

Fissile material.   Material that can sustain an explosive fission chain reaction—notably, highly enriched uranium or plutonium of any isotopic composition.

Fissile Material Cutoff Treaty (FMCT).   A proposed treaty to end the production of fissile material for weapons. It is sometimes also referred to as the Fissile Material (Cutoff) Treaty or Fissile Material Treaty.

Fission.   The process by which a nucleus or a heavy atom such as uranium or plutonium splits after absorbing a neutron or, in some cases, spontaneously. During the process of nuclear fission, typically two or three high-speed neutrons are emitted along with gamma rays. This is what makes a fission chain reaction possible.

Fissionable material.   A heavy isotope with an atomic nucleus that can undergo fission when struck by a neutron. Uranium-238 is a fissionable isotope in that it can be fissioned by high-energy neutrons. Unlike uranium-235, however, which is fissile as well as fissionable, uranium-238 cannot sustain a fission chain reaction.

Fission products.   Medium-weight isotopes such as krypton-90 and barium-144 that result from the fission of heavy isotopes. Most fission products are radioactive.

Fission weapon.   A nuclear weapon whose explosive energy is generated almost entirely by fission, in contrast to a thermonuclear weapon that generates a large fraction of its energy from fusion.

Gamma rays.   High-energy electromagnetic rays that carry off the extra energy released when a nucleus makes a transition from an excited to a lower energy state.

Gaseous diffusion.   A method of isotope separation based on the fact that gas molecules carrying isotopes with different masses diffuse through a porous barrier or membrane at different rates. Gaseous diffusion has been used to produce most of the global stockpile of HEU. Because the gas is pumped up to high pressure, it requires significant amounts of electric power.

Gigawatt (GW).   One billion watts. Used as a measure of the electrical (GWe) or thermal power (GWt) output of a large modern nuclear power plant. *See also* Megawatt (MW).

Glove box.   An enclosed working space kept at a slightly negative pressure with sleeves and gloves built in to allow workers to safely handle hazardous material.

Gun-type nuclear explosive.   A nuclear explosive in which a supercritical mass is created by firing one subcritical mass into another. This type of design works for highly enriched uranium but not plutonium. The Hiroshima bomb was a gun-type device.

Half-life. The time required for one-half of the nuclei in a quantity of a specific radioactive isotope to decay.

Heavy metal (HM). Typically used to describe the total mass of uranium and plutonium in reactor fuel. When used to characterize oxide fuels, the mass of heavy metal is the total fuel mass minus cladding and oxygen content.

Heavy water (D_2O). Water in which the abundant isotope of hydrogen, whose nucleus contains only one proton, is replaced by the rare isotope deuterium (0.01 percent of natural hydrogen), whose nucleus contains one proton and one neutron. It is used as a moderator to slow down neutrons in heavy water reactors.

Heavy water reactor (HWR). A reactor that uses heavy water as a neutron "moderator," to slow the neutrons between fissions. Heavy water is made by concentrating water molecules containing deuterium. Heavy water reactors typically use natural uranium as fuel. It is impossible to sustain a chain reaction in natural uranium in a reactor moderated by ordinary water because the "light" hydrogen in the water absorbs too many neutrons.

Highly enriched uranium (HEU). Uranium in which the percentage of uranium-235 nuclei has been increased from the natural level of 0.7 percent to 20 percent or more. A large fraction of HEU in the world as of 2013 was 90 percent enriched or higher because it was originally produced for weapon use.

Immobilization. Methods for the disposal of separated plutonium that involve mixing it with either glass ("vitrification") or ceramic-forming material. The resulting waste forms would be placed in a deep underground geological repository or a borehole.

Implosion-type nuclear explosive. A nuclear explosive in which a supercritical mass is created by compressing a subcritical mass to higher density. The Nagasaki bomb, whose core was plutonium, was an implosion-type device.

Improvised nuclear explosive device. A crude nuclear weapon assembled quickly using fissile material in forms that are on hand in a nuclear facility. It may be possible to assemble such a device using highly enriched uranium and achieve a kiloton-range nuclear yield. This would be much more difficult with plutonium.

Information Circulars (INFCIRCs). Records of note published by the International Atomic Energy Agency.

International Atomic Energy Agency (IAEA). An independent organization, established in 1957 under the United Nations, that is responsible for both promoting the peaceful use of nuclear technology and implementing "safeguards" agreements with non-weapon states under which it checks that fissile material is not diverted from peaceful uses or produced in undeclared facilities.

Isotope. A form of any element whose nucleus contains a specific number of neutrons. It is usually designated by the sum of the number of protons and neutrons in its nucleus (e.g., uranium-235 has 92 protons and 143 neutrons). Because all isotopes of an element have the same number of protons in the nucleus (92 for uranium) and therefore the same number of electrons, they have virtually the same chemical properties. But, because they have different numbers of neutrons in the nucleus, they have different atomic weights and nuclear properties. Uranium-235 can sustain a fission chain reaction, for example, while uranium-238, whose nucleus contains three more neutrons, cannot.

Kiloton (kt).   A unit used to measure the energy of a nuclear explosion, roughly the energy released by the explosion of one thousand tons of TNT, by definition equal to 4.184 terajoules (4.184×10^{12} joules). The fission of 1 kilogram of fissile material releases about 18 kilotons of TNT equivalent.

Krypton-85.   A radioactive isotope of the element krypton, an inert gas, produced by the fission of uranium and plutonium. It has a half-life of 10.7 years. It accumulates in nuclear fuel during irradiation and usually is released to the atmosphere during reprocessing.

Light water.   Ordinary water (H_2O) as distinguished from heavy water (D_2O), which contains deuterium, a heavier isotope of hydrogen.

Light water reactor (LWR).   A reactor that uses ordinary water to cool the reactor core and to "moderate" the speed of neutrons between fissions and increase the probability of their capture in uranium-235. LWRs usually use low-enriched uranium as fuel. It is the most common nuclear power reactor design. It was originally developed to power nuclear submarines.

Low-enriched uranium (LEU).   Uranium in which the percentage of uranium-235 nuclei has been increased from the natural level of 0.7 percent to less than 20 percent. The fuel of light water reactors is usually enriched to 4–5 percent.

Magnox reactor.   A natural uranium-fueled, graphite-moderated, carbon dioxide–cooled nuclear reactor designed and widely used in the United Kingdom. It was the reactor design used in the world's first commercial nuclear power plant, which came online in 1956. The United Kingdom also used such reactors for producing weapon-grade plutonium. These reactors are now being retired.

Megawatt (MW).   One million watts. Used as a measure of electrical power output of a nuclear power plant: 1 million watts of electric power (megawatts electric, or MWe). Also used to measure the rate at which heat is released in research or plutonium production reactors: 1 million watts of thermal energy (megawatts thermal, or MWt). A typical light water power reactor today has a peak electricity generation capacity of approximately 1,000 MWe—that is, 10^9 watts. Such a reactor would generate about 3,000 MWt.

Megawatt-day (MWt-day).   A unit of energy. The cumulative amount of heat that would be released in a day by a reactor producing heat at a rate of one megawatt. The fission of one gram of uranium or plutonium releases approximately one MWt-day of thermal energy.

Mixed uranium plutonium oxide (MOX) fuel.   Nuclear reactor fuel composed of a mixture of plutonium and natural or depleted uranium in oxide form, commonly referred to as MOX fuel. The plutonium replaces the uranium-235 in low-enriched uranium as the primary fissioning material in the fuel. MOX is used in Europe—and its use is planned in India and Japan—to recycle plutonium recovered from spent fuel through reprocessing.

Natural uranium.   Uranium as found in nature, containing 0.7 percent of uranium-235, 99.3 percent of uranium-238, and trace quantities of uranium-234 (formed by the radioactive decay of uranium-238 via thorium-234 and protactinium-234).

Neptunium-237 (Np-237).   A two-million-year half-life fissile isotope produced in nuclear reactors by two successive neutron captures on uranium-235 and a radioactive decay. There is no public information that Np-237 has actually ever been used in a nuclear weapon, but its properties make it as suitable as uranium-235. The IAEA considers it an "alternative nuclear material."

Neutron.   An uncharged elementary particle with a mass slightly greater than that of a proton. Neutrons are found in the nuclei of every atom heavier than hydrogen. Neutrons provide the links in a fission chain reaction.

New START.   The 2010 treaty between the United States and Russian Federation to reduce and limit their deployed strategic nuclear forces. The treaty permits each country to have 700 deployed strategic nuclear weapon delivery systems (long-range missiles and nuclear bombers) and caps the total number of warheads that they carry (missiles) or are equipped to carry (bombers) to 1,550. The treaty entered into force in 2011 and its limits are to be met no later than 2018. Verification includes reciprocal inspections of missile and bomber bases. *See also* Strategic Arms Reduction Theory (START).

Non-Proliferation Treaty (NPT).   The Treaty on the Non-Proliferation of Nuclear Weapons, which entered into force in 1970. The NPT aims to (1) prevent the spread of nuclear weapons to states other than those five that had tested nuclear weapons by 1967, (2) foster the peaceful uses of nuclear energy under international safeguards in non-weapon states, and (3) further the goal of nuclear disarmament.

Nuclear fuel.   The most commonly used nuclear fuels are low-enriched and natural uranium. Highly enriched uranium and mixed oxide fuel are also used to fuel some reactors.

Nuclear fuel cycle.   The chemical and physical operations needed to prepare nuclear material for use in reactors and to dispose of or recycle the material after its removal from the reactor. Existing fuel cycles begin with the mining of uranium ore and produce fissile plutonium as a by-product by absorption of neutrons in uranium-238 while the fuel is in the reactor. Some proposed fuel cycles would use natural thorium as a fertile material to produce the fissile isotope uranium-233, which would then be recycled in reactor fuel. An "open" fuel cycle stores the spent fuel indefinitely. A "closed" fuel cycle reprocesses it and recycles the fissile and fertile material once or more and stores the fission products and other radioactive isotopes.

Nuclear reactor.   An arrangement of nuclear and other materials designed to sustain a controlled nuclear chain reaction that releases heat. Nuclear reactors fall into three general categories: power and propulsion reactors; production reactors (for producing plutonium, tritium, and also radioactive isotopes used in medicine); and research reactors. The heat generated by a power or propulsion reactor is converted into electrical or mechanical power. Most reactors produce plutonium in their irradiated fuel.

Nuclear Suppliers Group (NSG).   A group of nuclear technology and nuclear material exporting countries organized in 1977 with agreed export guidelines. The guidelines currently include a "trigger list" of items that the suppliers agree to export only to a nonnuclear weapon state that is a party to the NPT or a state outside the NPT if that state has an agreement with the IAEA that allows the agency to safeguard all its nuclear activities. In 2008, at U.S. insistence, an exception was made for India.

Nuclear waste.   Spent nuclear fuel is highly radioactive and, in most countries, considered a form of waste. Reprocessing spent fuel yields other waste streams, including fission products and transuranic elements. These waste forms are stored pending final disposal.

Pit.   A solid mass or hollow shell of fissile material (usually plutonium, sometimes HEU, or a composite of both) clad by a protective metal such as steel. In a nuclear weapon, the pit is surrounded by high explosive that, when triggered, compresses the fissile material into a supercritical state where it can undergo an explosive chain reaction.

Plutonium-239 (Pu-239).   A fissile isotope with a half-life of about 24,000 years produced when uranium-238 captures an extra neutron. The plutonium used in the core of nuclear weapons typically contains more than 90 percent Pu-239 and is described as "weapon-grade."

Plutonium-240 (Pu-240).   An isotope with a half-life of 6,600 years produced in reactors when a plutonium-239 nucleus absorbs a neutron without fissioning. Its high rate of neutron emission from spontaneous fission makes it undesirable in weapon plutonium.

Plutonium-241 (Pu-241).   A fissile isotope with a half-life of fourteen years produced in reactors by neutron absorption on plutonium-240. Pu-241 decays into americium-241.

Plutonium Uranium Extraction (PUREX).   Method for separating plutonium from spent nuclear fuel. *See also* Reprocessing.

Power reactor.   A reactor whose purpose is to produce heat to generate electricity—usually by generating high-pressure steam that drives a turbine.

Production reactor.   A reactor designed primarily for the large-scale production of plutonium for weapons and/or tritium. Some production reactors have been dual purpose, generating power as a by-product.

Radioactivity.   The spontaneous disintegration of an unstable atomic nucleus resulting in the emission of electrons (beta decay) or helium nuclei (alpha decay). Often the new nucleus is produced in an "excited" state that emits its excess energy in the form of gamma rays (high-energy X-rays).

Reactor-grade plutonium.   The United States defines reactor-grade plutonium as containing more than 18 percent plutonium-240—much more than in weapon-grade plutonium. Reactor-grade plutonium can be used, however, to make a nuclear explosive.

Reprocessing.   The chemical treatment of spent reactor fuel to separate plutonium and uranium from fission products. Because of the intense radioactivity of the fission products, this has to be done remotely behind heavy shielding.

Research reactor.   A reactor designed primarily to supply neutron irradiation for experimental purposes. It also may be used for training, the testing of reactor materials, and the production of radioisotopes.

Safeguards.   Measures aimed at detecting in a timely fashion the diversion of significant quantities of fissile material from monitored, peaceful, nuclear activities. For nonnuclear weapon states that are parties to the Non-Proliferation Treaty, the safeguards are implemented by the IAEA. *See also* Significant quantity (SQ).

Separative work unit (SWU).   A measure of the amount of work that must be performed on an isotopic mixture of a given composition to produce from this feed material a specified amount of product enriched in a given isotope and of depleted "tails." For a uranium enrichment plant, the capacity is given typically in kg-SWU/year. A 1 GWe light water nuclear power reactor requires about 100,000 kg-SWU of separative work to produce from natural uranium one year's fuel supply of 20 tons of uranium enriched to 4 percent uranium-235. This could be provided by 10,000 centrifuges each with rating of 20 kg-SWU/year, operating for six months.

Significant quantity (SQ).   A term used by the IAEA for the amount of fissile material required to manufacture a first-generation nuclear explosive device. This quantity includes expected losses during manufacturing. In setting its inspection goals, the IAEA assumes these quantities to be: 8 kilograms of plutonium containing less than 80 percent plutonium-238, 8 kilograms of uranium-233, and 25 kilograms of uranium-235 in highly enriched uranium.

Spent fuel.   Fuel elements that have been removed from a reactor usually because the fissionable material they contain has been depleted to a level near where it can no longer sustain a chain reaction. The high concentration of radioactive fission products in spent power reactor fuel creates a gamma radiation field around it that makes spent light water reactor fuel "self-protecting" for about one hundred years. A few years after discharge, the gamma field at a distance of a meter would be lethal in minutes. A century after discharge it would be lethal in a few hours.

Strategic Arms Reduction Treaty (START).   The 1991 START I treaty, which limited the United States and Russia to 1,600 strategic nuclear weapon delivery systems (long-range missiles and bombers) each and capped the total number of warheads that they carried (missiles) or are equipped to carry (bombers). It expired in 2009 and was replaced by the New START Treaty, with lower limits.

Thermonuclear explosive.   A type of nuclear weapon, also known as a hydrogen bomb, that produces much of its energy through nuclear fusion reactions of the heavy hydrogen isotopes deuterium and tritium. These fusion reactions require temperatures around 100 million degrees centigrade created by a fission explosive "trigger." Thermonuclear weapons can have yields much larger than simple fission weapons.

Thorium-232 (Th-232). The naturally occurring isotope of thorium. It can be used to breed uranium-233 when irradiated in a nuclear reactor. After neutron absorption, it becomes thorium-233, which decays into the fissile isotope uranium-233.

Transuranic. Any element whose atomic number is higher than that of uranium. All transuranics are produced artificially and are radioactive. The most commonly produced transuranic isotopes, in order of increasing weight, are neptunium, plutonium, americium, and curium.

Tritium. The heaviest hydrogen isotope, containing one proton and two neutrons in its nucleus. It has a half-life of 12.3 years and is produced in reactors and in thermonuclear weapons by splitting lithium-6 with a neutron. *See also* Boosting.

Uranium. The element with 92 protons and electrons. It is used in a variety of forms in the nuclear fuel cycle and in nuclear weapons. *See also* Highly enriched uranium (HEU), Low-enriched uranium (LEU), and Natural uranium.

Uranium dioxide (UO_2). The chemical form of uranium used in heavy water and light water power reactor fuel. Powdered uranium dioxide is pressed and then sintered into ceramic fuel pellets.

Uranium hexafluoride (UF_6). A volatile compound of uranium and fluorine. UF_6 is a solid at atmospheric pressure and room temperature, but can be transformed into gas by heating. UF_6 gas is the feedstock in gas centrifuge and gaseous diffusion uranium enrichment processes.

Uranium oxide (U_3O_8). The most common oxide of uranium found in typical ores. Uranium oxide is extracted from the ore during the milling process. The ore may contain only 0.1 percent uranium. Yellowcake, the product of the milling process, contains over 80 percent uranium.

Uranium-233 (U-233). An artificial fissile isotope produced by neutron absorption in thorium-232. Like HEU and plutonium, it is weapon-usable. It has been used in at least one nuclear test but not in deployed nuclear weapons. Uranium-233 has been of interest as a reactor fuel for heavy and light water moderated reactors because its fission by low-energy neutrons releases more neutrons than does the fission of plutonium-239. Along with uranium-233, minute quantities of uranium-232 are generally also produced. A decay product of uranium-232 produces high energy and penetrating gamma radiation resulting in a stronger radiation field than from uranium-235, uranium-238, or plutonium.

Uranium-235 (U-235). The only naturally occurring chain-reacting isotope. Natural uranium contains 0.7 percent U-235. Light-water reactors use fuel containing 4–5 percent U-235. Weapon-grade uranium normally contains at least 90 percent.

Uranium-238 (U-238). Natural uranium contains approximately 99.3 percent U-238. It can be used to breed plutonium when irradiated in a nuclear reactor. After neutron absorption, it becomes uranium-239, which decays via neptunium-239 into the fissile isotope plutonium-239.

Vitrification.   The immobilization of liquid high-level radioactive waste by mixing the material into glass for long-term storage or final disposal.

Weapon-grade.   Fissile material with the isotopic makeup typically used in fission explosives: uranium enriched to over 90 percent uranium-235 or plutonium that is more than 90 percent plutonium-239. Uranium enriched to greater than 20 percent and any isotopic mixture of plutonium containing less than 80 percent plutonium-238 are not weapon-grade but are considered weapon-usable. The uranium used in the Hiroshima weapon was enriched to an average of about 80 percent.

Yellowcake.   A uranium concentrate produced during the process of extracting uranium from ore that contains over 80 percent U_3O_8. In preparation for uranium enrichment, the yellowcake is converted to UF_6. In the preparation of natural uranium fuel for heavy water reactors, yellowcake is processed into uranium metal or uranium dioxide (UO_2).

Yield.   The total energy released in a nuclear explosion—usually measured by the number of kilotons of TNT whose explosion would release the same amount of energy. *See also* Kiloton (kt).

Bibliography

2000 Review Conference of the Parties to the Treaty on the Non-Proliferation of Nuclear Weapons, Final Document, Parts I and II. NPT/CONF.2000/28. New York: United Nations, 2000.

2010 Review Conference of the Parties to the Treaty on the Non-Proliferation of Nuclear Weapons, Final Document, Volume 1. New York: United Nations, 2010.

Abraham, Itty. *The Making of the Indian Atomic Bomb: Science, Secrecy and the Postcolonial State*. New York: Zed Books, 1998.

Additional Information Concerning Underground Nuclear Weapon Test of Reactor-Grade Plutonium. Washington, DC: U.S. Department of Energy, Office of the Press Secretary, June 27, 1994.

"Agreement between the Government of the United States of America and the Government of the Russian Federation Concerning the Disposition of Highly Enriched Uranium Extracted from Nuclear Weapons," February 18, 1993. www.fissilematerials.org/library/heu93.pdf.

"Agreement between the Government of the United States of America and the Government of the Russian Federation Concerning the Management and Disposition of Plutonium Designated as No Longer Required for Defense Purposes and Related Cooperation," 2000. www.fissilematerials.org/library/gov00.pdf.

"Agreement between the Government of the United States of America and the Government of the United Kingdom of Great Britain and Northern Ireland for Cooperation on the Uses of Atomic Energy for Mutual Defense Purposes." London: Her Majesty's Stationery Office, 1958.

Aide Memoire of Conversation between the President and the Prime Minister at Hyde Park, September 18, 1944, in U.S. Department of State, Foreign Relations of the United States: Conference at Quebec, 1944. Washington, DC: U.S. Government Printing Office, 1944.

Albright, David. *Peddling Peril: How the Secret Nuclear Trade Arms America's Enemies*. New York: Free Press, 2010.

Albright, David, and Robert Avagyan. *Construction Progressing Rapidly on the Fourth Heavy Water Reactor at the Khushab Nuclear Site*. Washington, DC: Institute for Science and International Security, May 21, 2012.

Albright, David, Frans Berkhout, and William Walker. *Plutonium and Highly Enriched Uranium 1996: World Inventories, Capabilities and Policies.* Oxford: Oxford University Press, 1997.

Albright, David, and Paul Brannan. *Chashma Nuclear Site in Pakistan with Possible Reprocessing Plant.* Washington, DC: Institute for Science and International Security, January 2007.

Albright, David, and Paul Brannan. *Pakistan Expanding Plutonium Separation Facility Near Rawalpindi.* Washington, DC: Institute for Science and International Security, May 19, 2009.

Albright, David, and Harold A. Feiveson. "Plutonium Recycling and the Problem of Nuclear Proliferation." *Annual Review of Energy* 13 (1988): 239–265.

Albright, David, and Kimberly Kramer. *Neptunium 237 and Americium: World Inventories and Proliferation Concerns (Revised).* Washington, DC: Institute for Science and International Security, August 22, 2005.

Albright, David, and Kevin O'Neill. *Solving the North Korean Nuclear Puzzle.* Washington, DC: Institute for Science and International Security, 2000.

Alvarez, Luis W. *Adventures of a Physicist.* New York: Basic Books, 1987.

Analysis of Uranium Supply to 2050. 2001. Austria: International Atomic Energy Agency.

Andrews, Anthony. *Nuclear Fuel Reprocessing: U.S. Policy Development.* RS22542. Washington, DC: Congressional Research Service, Library of Congress, March 27, 2008.

Arms, Nancy. *A Prophet in Two Countries: The Life of F. E. Simon.* Oxford, New York: Pergamon Press, 1966.

Arnold, Lorna. *Windscale, 1957: Anatomy of a Nuclear Accident.* New York: St. Martin's Press, 1992.

Assessing the Options: Future Management of Used Nuclear Fuel in Canada. 2004. Canada: Nuclear Waste Management Organization.

Banning the Production of Fissile Materials for Nuclear Weapons: Country Perspectives on the Challenges to a Fissile Material (Cutoff) Treaty. International Panel on Fissile Materials, September 2008. www.fissilematerials.org/library/gfmr08cv.pdf.

Bari, R., L. Y. Cheng, J. Phillips, J. Pilat, G. Rochau, I. Therios, R. Wigeland, E. Wonder, and M. Zentner. *Proliferation Risk Reduction Study of Alternative Spent Fuel Processing.* BNL-90264-2009-CP. Upton, NY: Brookhaven National Laboratory, July 2009.

Barnaby, Frank, ed. *Nuclear Energy and Nuclear Weapon Proliferation.* London: Taylor and Francis, 1979.

Barnard, C. I., J. R. Oppenheimer, C. A. Thomas, H. A. Winne, and D. E. Lilienthal. *A Report on the International Control of Atomic Energy.* Washington, DC, 1946. www.fissilematerials.org/library/ach46.pdf.

Benedict, Manson, Thomas H. Pigford, and Hans Wolfgang Levi. *Nuclear Chemical Engineering*. 2nd ed. New York: McGraw-Hill Book Company, 1981.

Berkhout, Frans, Anatoli Diakov, Harold Feiveson, Helen Hunt, Edwin Lyman, Marvin Miller, and Frank von Hippel. "Disposition of Separated Plutonium." *Science & Global Security* 3, nos. 3–4 (1993): 161–213.

Betts, Richard K. *Nuclear Blackmail and Nuclear Balance*. Washington, DC: Brookings Institution, 1987.

Birch, A. Francis. *Report of Gun Assembled Nuclear Bomb*, October 6, 1945.

Bohr, Niels, and John A. Wheeler. "The Mechanism of Nuclear Fission." *Physical Review* 56, no. 5 (1939): 426–450.

Bothwell, Robert. *Nucleus: The History of Atomic Energy of Canada Limited*. Toronto: University of Toronto Press, 1988.

Bradley, Donald J. *Behind the Nuclear Curtain: Radioactive Waste Management in the Former Soviet Union*. Columbus: Pacific Northwest National Laboratory/Batelle Press, 1997.

Brady, P. V., Bill W. Arnold, Susan Altman, and Peter Vaughn. *Deep Borehole Disposal of High-Level Radioactive Waste*. SAND2012-7789. Albuquerque, NM: Sandia National Laboratories, 2012.

Brady, P. V., and M. J. Driscoll. "Deep Borehole Disposal of Nuclear Waste." *Radwaste Solutions* 17, no. 5 (September/October 2010): 58.

Brian, Danielle, and Peter Stockton. *U.S. Nuclear Weapons Complex: Security at Risk*. Washington, DC: Project on Government Oversight, 2001.

Bukharin, Oleg. "Analysis of the Size and Quality of Uranium Inventories in Russia." *Science & Global Security* 6 (1996): 59–77.

Bukharin, Oleg, and William Potter. "Potatoes Were Guarded Better." *Bulletin of the Atomic Scientists* 51, no. 3 (May/June 1995): 46–50.

Bundy, McGeorge. *Danger and Survival: Choices about the Bomb in the First Fifty Years*. New York: Random House, 1988.

Bustamante, Jacqueline, et al. *Disposition of Fuel Elements from the Aberdeen and Sandia Pulse Reactor (SPR-II) Assemblies*. LA-UR-10-03828. Los Alamos, NM: Los Alamos National Laboratory, 2010.

Campbell, Kurt M., Robert J. Einhorn, and Mitchell Reiss. *The Nuclear Tipping Point: Why States Reconsider Their Nuclear Choices*. Washington, DC: Brookings Institution Press, 2004.

Cardis, Elisabeth, Daniel Krewski, Mathieu Boniol, Vladimir Drozdovitch, Sarah C. Darby, Ethel S. Gilbert, Suminori Akiba, et al. "Estimates of the Cancer Burden in Europe from Radioactive Fallout from the Chernobyl Accident." *International Journal of Cancer* 119, no. 6 (2006): 1224–1235.

Carlisle, Rodney P., and Joan M. Zenzen. *Supplying the Nuclear Arsenal: American Production Reactors, 1942–1992*. Baltimore, MD: Johns Hopkins University Press, 1996.

Chapman, Neil, and Fergus Gibb. "A Truly Final Waste Management Solution: Is Very Deep Borehole Disposal a Realistic Option for High-Level Waste or Fissile Materials?" *Radwaste Solutions* 10, no. 4 (July/August 2003): 26–37.

Charpin, J. M., B. Dessus, and R. Pellat. *Economic Forecast Study of the Nuclear Power Option, Report to the Prime Minister*, Paris, July 2000.

Choice of Method—Evaluation of Strategies and Systems for Disposal of Spent Nuclear Fuel. P-10-47. Stockholm: SKB, October 2010.

Clark, David L., and David E. Hobart. "Reflections on the Legacy of a Legend: Glenn T. Seaborg, 1912–1999." *Los Alamos Science*, no. 26 (2000): 56–61.

Clements, Tom, Edwin Lyman, and Frank von Hippel. "The Future of Plutonium Disposition," *Arms Control Today*, July/August 2013, 8–15.

Clark, Grenville, and Louis B. Sohn. *World Peace through World Law*. 2nd ed. Piscataway, NJ: Transaction Publishers, 1960.

Cochran, Thomas B. "Highly Enriched Uranium Production for South African Nuclear Weapons." *Science & Global Security* 4, no. 1 (1994): 161–176.

Cochran, Thomas B., William M. Arkin, and Milton M. Hoenig. *U.S. Nuclear Forces and Capabilities, Vol. I: Nuclear Weapons Databook*. Cambridge, MA: Ballinger Publishing Company, 1984.

Cochran, Thomas B., Harold A. Feiveson, Walt Patterson, Gennadi Pshakin, M. V. Ramana, Mycle Schneider, Tatsujiro Suzuki, and Frank von Hippel. *Fast Breeder Reactor Programs: History and Status*. Princeton, NJ: International Panel on Fissile Materials, February 2010. www.fissilematerials.org/library/rr08.pdf.

Cochran, Thomas B., Robert S. Norris, and Oleg Bukharin. *Making the Russian Bomb: From Stalin to Yeltsin*. 1995. Boulder, CO: Westview Press.

Committee on America's Energy Future, National Academy of Sciences, National Academy of Engineering; National Research Council. *America's Energy Future: Technology and Transformation: Summary Edition*. Washington, DC: The National Academies Press, 2009.

Committee on International Security and Arms Control. *Management and Disposition of Excess Weapons Plutonium: Reactor-Related Options*. Washington, DC: U.S. National Academy of Sciences, 1995.

Committee on International Security and Arms Control. *Management and Disposition of Excess Weapons Plutonium*. Washington, DC: U.S. National Academy of Sciences, 1994.

Committee on Separations Technology and Transmutation Systems. *Nuclear Wastes: Technologies for Separations and Transmutation*. Washington, DC: National Academy Press, 1996.

Comparative Analysis of Alternative Financing Plans for the Clinch River Breeder Reactor Project. Washington, DC: U.S. Congressional Budget Office, September 1983.

Convention on the Prohibition of the Development, Production, Stockpiling and Use of Chemical Weapons and on Their Destruction (Chemical Weapons

Convention). The Hague, the Netherlands: Organization for the Prohibition of Chemical Weapons, 2005.

Cost Analysis and Economics in Plutonium Disposition: Cost Estimates for the Disposition of Weapon-Grade Plutonium Withdrawn from Military Programs. Washington, DC: Joint U.S.-Russian Working Group on Cost Analysis and Economics in Plutonium Disposition, 2001.

Coster-Mullen, John. *Atom Bombs: The Top Secret Inside Story of Little Boy and Fat Man.* Self-published, 2005.

De Geer, Lars-Erik. "Radionuclide Evidence for Low-Yield Nuclear Testing in North Korea in April/May 2010." *Science & Global Security* 20 (2012): 1–29.

Declassification of Today's Highly Enriched Uranium Inventories at Department of Energy Laboratories. Washington, DC: U.S. Department of Energy, Office of the Press Secretary, June 27, 1994. www.fissilematerials.org/library/doe06a.pdf.

Difficulties in Determining if Nuclear Training of Foreigners Contributes to Weapons Proliferation, Report by the Comptroller General of the United States. Washington, DC: General Accounting Office, April 23, 1979.

Dingman, Roger. "Atomic Diplomacy during the Korean War." *International Security* 13, no. 3 (1989): 50–91.

Director, Naval Nuclear Propulsion. *Report on Use of Low Enriched Uranium in Naval Nuclear Propulsion.* Washington, DC: U.S. Department of Energy, June 1995. www.fissilematerials.org/library/onnp95.pdf.

Draft Surplus Plutonium Supplemental Environmental Impact Statement. DOE/ EIS00283–S2. Washington, DC: U.S. Department of Energy, 2012.

Eisenhower, Dwight D. "Peaceful Uses of Atomic Energy." Presented at the General Assembly of the United Nations, New York, December 8, 1953.

ElBaradei, Mohamed. "Towards a Safer World." *The Economist,* October 16, 2003: 47–48.

Energy, Electricity and Nuclear Power Estimates for the Period up to 2050, 2012 Edition. Vienna: International Atomic Energy Agency, 2013.

Fast Breeders, International Nuclear Fuel Cycle Evaluation (INFCE). Vienna: International Atomic Energy Agency, 1980.

Feiveson, Harold A., ed. *The Nuclear Turning Point: A Blueprint for Deep Cuts and De-alerting of Nuclear Weapons.* Washington, DC: Brookings Institution, 1999.

Feiveson, Harold A. "Latent Proliferation: The International Security Implications of Civilian Nuclear Power." PhD thesis, Princeton University, 1972.

Feiveson, Harold A., Zia Mian, M. V. Ramana, and Frank von Hippel, eds. *Managing Spent Fuel from Power Reactors: Experience and Lessons from Around the World.* Princeton, NJ: International Panel on Fissile Materials, September 2011.

Feld, Bernard T., Gertrud Weiss Szilard, Jaques Monod, Carl Eckart, and Kathleen R. Winsor. *The Collected Works of Leo Szilard.* Cambridge, MA: MIT Press, 1972.

Ferguson, D. E. "Simple, Quick Processing Plant." Intra-Laboratory Correspondence. Oak Ridge, TN: Oak Ridge National Laboratory, August 30, 1977.

Fetter, Steve. "Nuclear Archaeology: Verifying Declarations of Fissile-Material Production." *Science & Global Security* 3, nos. 3–4 (1993): 237–259.

Fetter, Steve, and Frank von Hippel. "Is U.S. Reprocessing Worth The Risk?" *Arms Control Today* 35, no. 7 (2005): 6–12.

Fetter, Steve, and Frank von Hippel. "A Step-by-Step Approach to a Global Fissile Materials Cutoff." *Arms Control Today* 25, no. 8 (1995): 3–8.

Fetter, Steve, and Frank von Hippel. "The Hazard from Plutonium Dispersal by Nuclear-Warhead Accidents." *Science & Global Security* 2 (1990): 21–41.

Fischer, David. *History of the International Atomic Energy Agency: The First Forty Years*. Vienna: IAEA, 1997.

Fitts, R. B., and H. Fujii. "Fuel Cycle Demand, Supply and Cost Trends." *IAEA Bulletin* 18, no. 1 (1976): 19–24.

Forwood, Martin. *The Legacy of Reprocessing in the United Kingdom*. International Panel on Fissile Materials, July 2008.

Franck, James, Donald J. Hughes, J. J. Nickson, Eugene Rabinowitch, Glenn T. Seaborg, J. C. Stearns, and Leo Szilard. *Report of the Committee on Political and Social Problems Manhattan Project* (The Franck Report). University of Chicago, June 11, 1945. www.fissilematerials.org/library/fra45.pdf.

Fréchou, C., and D. Calmet. "^{129}I in the Environment of the La Hague Nuclear Fuel Reprocessing Plant—from Sea to Land." *Journal of Environmental Radioactivity* 70, nos. 1–2 (2003): 43–59.

Frisch, Otto, and Rudolph Peierls. "Memorandum on the Properties of a Radioactive 'Super-bomb.'" Birmingham, UK: Birmingham University, March 1940.

Frisch, Otto, and Rudolph Peierls. "On the Construction of a Super-bomb Based on a Nuclear Chain Reaction in Uranium, Memorandum." Birmingham, UK: Birmingham University, March 1940.

Garland, J. A., and R. Wakeford. "Atmospheric Emissions from the Windscale Accident of October 1957." *Atmospheric Environment* 41, no. 18 (2007): 3904–3920.

Gasner, Alex, and Alexander Glaser. "Nuclear Archaeology for Heavy-Water-Moderated Plutonium Production Reactors." *Science & Global Security* 19, no. 3 (2011): 223–233.

General and Complete Disarmament. Resolution A/RES/48/75. New York: United Nations General Assembly, December 16, 1993. www.un.org/documents/ga/res/48/a48r075.htm.

Gephart, R. E. *A Short History of Hanford Waste Generation, Storage, and Release*. PNNL-13605 Rev. 4. Richland, WA: Pacific Northwest National Laboratory, October 2003.

Gilinsky, Victor, and Roger J. Mattson. "Revisiting the NUMEC Affair." *Bulletin of the Atomic Scientists* 66, no. 2 (April 2010): 61–75.

Gilinsky, Victor, Marvin Miller, and Harmon Hubbard. *A Fresh Examination of the Proliferation Dangers of Light Water Reactors.* Washington, DC: The Nonproliferation Policy Education Center, October 22, 2004.

Gilpin, Robert. *American Scientists and Nuclear Weapons Policy.* Princeton, NJ: Princeton University Press, 1962.

Glaser, Alexander. "Beyond A. Q. Khan: The Gas Centrifuge, Nuclear Weapon Proliferation, and the NPT Regime." *INESAP Information Bulletin* 23 (2004): 1–5.

Glaser, Alexander. "Characteristics of the Gas Centrifuge for Uranium Enrichment and Their Relevance for Nuclear Weapon Proliferation." *Science & Global Security* 16, nos. 1–2 (2008): 1–25.

Glaser, Alexander. *Internationalization of the Nuclear Fuel Cycle.* ICNND Research Paper No. 9. Canberra, Australia: International Commission on Nuclear Non-proliferation and Disarmament, February 2009. www.icnnd.org.

Glaser, A., and S. Bürger. "Verification of a Fissile Material Cutoff Treaty: The Case of Enrichment Facilities and the Role of Ultra-trace Level Isotope Ratio Analysis." *Journal of Radioanalytical and Nuclear Chemistry* 280, no. 1 (April 2009): 85–90.

Glaser, Alexander, and Uwe Filges. "Neutron-Use Optimization with Virtual Experiments to Facilitate Research-Reactor Conversion to Low-Enriched Fuel." *Science & Global Security* 20, nos. 2–3 (2012): 141–154.

Glaser, Alexander, and Marvin Miller. *Estimating Plutonium Production at Israel's Dimona Reactor.* CA: Palm Desert, 2011.

Glaser, Alexander, and M. V. Ramana. "Weapon-Grade Plutonium Production Potential in the Indian Prototype Fast Breeder Reactor." *Science & Global Security* 15, no. 2 (2007): 85–105.

Global Fissile Material Report 2007. Princeton, NJ: International Panel on Fissile Materials, September 2007. www.fissilematerials.org/library/gfmr07.pdf.

Global Fissile Material Report 2008: Scope and Verification of a Fissile Material (Cutoff) Treaty. Princeton, NJ: International Panel on Fissile Materials, 2008. www.fissilematerials.org/library/gfmr08.pdf.

Global Fissile Material Report 2009: A Path to Nuclear Disarmament. Princeton, NJ: International Panel on Fissile Materials, October 2009. www.fissilematerials.org/library/gfmr09.pdf.

Global Fissile Material Report 2010: Balancing the Books: Production and Stocks. Princeton, NJ: International Panel on Fissile Materials, 2010. www.fissilematerials.org/library/gfmr10.pdf.

Global Fissile Material Report 2011: Nuclear Weapon and Fissile Material Stockpiles and Production. Princeton, NJ: International Panel on Fissile Materials, January 2012. www.fissilematerials.org/library/gfmr11.pdf.

Global Fissile Material Report 2013: Increasing Transparency of Nuclear Warhead and Fissile Material Stocks as a Step toward Disarmament. Princeton, NJ: International Panel on Fissile Materials, November 2013. www.fissilematerials.org/library/gfmr13.pdf.

Goldschmidt, Bertrand. *The Atomic Complex: A Worldwide Political History of Nuclear Energy.* La Grange Park, IL: American Nuclear Society, 1982.

Goldston, Robert J., and Alexander Glaser. "Inertial Confinement Fusion Energy R&D and Nuclear Proliferation: The Need for Direct and Transparent Review." *Bulletin of the Atomic Scientists* 67, no. 3 (May 2011): 59–66.

Gordin, Michael D. *Red Cloud at Dawn: Truman, Stalin, and the End of the Atomic Monopoly.* New York: Farrar, Straus and Giroux, 2009.

Gowing, Margaret. *Britain and Atomic Energy 1939–1945.* London: Macmillan and Co., 1964.

Gowing, Margaret, and Lorna Arnold. *Independence and Deterrence: Britain and Atomic Energy, 1945–1952.* New York: St. Martin's Press, 1974.

HAL [High Active Liquor] Stocks, Specification No. 793, Project Assessment Report. Bootle, Merseyside, UK: UK Health and Safety Executive, Office of for Nuclear Regulation, 2011, www.hse.gov.uk.

Handwerk, J. H., and R. C. Lied. *The Manufacture of the Graphite-Urania Fuel Matrix for TREAT.* ANL-5963. Argonne, IL: Argonne National Laboratory, 1960.

Hecker, Siegfried S. *A Return Trip to North Korea's Yongbyon Nuclear Complex.* Stanford, CA: Center for International Security and Cooperation, Stanford University, November 20, 2010.

Hersh, Seymour M. *The Samson Option: Israel's Nuclear Arsenal and American Foreign Policy.* New York: Random House, 1991.

Hewlett, Richard G., and Oscar E. Anderson Jr. *The New World: A History of the United States Atomic Energy Commission, Volume 1, 1939–1946.* California Studies in the History of Science. Berkeley: University of California Press, 1962.

Hewlett, Richard G., and Francis Duncan. *Nuclear Navy, 1946–1962.* Chicago: University of Chicago Press, 1974.

Hibbs, Mark. "After 30 Years, PAEC Fulfills Munir Khan's Plutonium Ambition." *Nucleonics Week*, June 15, 2000, 13.

Hibbs, Mark. "Bhutto May Finish Plutonium Reactor without Agreement on Fissile Stocks." *Nucleonics Week*, October 6, 1994, 10.

"High Levels of Radioactive Carbon Contamination around La Hague." Greenpeace press release, Paris, November 12, 1998.

Highly Enriched Uranium: Striking a Balance. A Historical Report on the United States Highly Enriched Uranium Production, Acquisition, and Utilization Activities from 1945 through September 30, 1996. Washington, DC: U.S. Department of Energy, December 2005. www.fissilematerials.org/library/doe01rev.pdf.

Highly Enriched Uranium Inventory: Amounts of Highly Enriched Uranium in the United States. Washington, DC: U.S. Department of Energy, 2006. www .fissilematerials.org/library/doe06f.pdf.

Hinton, J. P., R. W. Barnard, D. E. Bennett, R. W. Crocker, M. J. Davis, H. J. Groh, E. A. Hakkila, G. A. Harms, W. L. Hawkins, E. E. Hill, L. W. Kruse, J. A. Milloy, W. A. Swansiger, K. J. Ystesund, et al. *Proliferation Resistance of Fissile Material Disposition Program Plutonium Disposition Alternatives: Report of the Proliferation Vulnerability Red Team.* SAND97-8201. Albuquerque, NM: Sandia National Laboratories, October 1996. www.fissilematerials.org/library/ doe96c.pdf.

Historical Accounting for UK Defence Highly Enriched Uranium. London: UK Ministry of Defence, March 2006. www.fissilematerials.org/library/mod06.pdf.

Hoddeson, Lillian, Paul W. Henriksen, Roger A. Meade, and Catherine Westfall. *Critical Assembly: A Technical History of Los Alamos during the Oppenheimer Years, 1943–1945.* Cambridge: Cambridge University Press, 1993.

Holloway, David. *Stalin and the Bomb.* New Haven: Yale University Press, 1994.

Hurst, Donald G. *Canada Enters the Nuclear Age: a Technical History of Atomic Energy of Canada Limited.* Montreal: McGill-Queen's University Press, 1997.

IAEA. *Non-HEU Production Technologies for Molybdenum-99 and Technetium-99m.* IAEA Nuclear Energy Series NF-T-5.4. Vienna: International Atomic Energy Agency, 2013.

IAEA Annual Report 2011. Vienna: International Atomic Energy Agency, 2012.

IAEA Annual Report 2012. Vienna: International Atomic Energy Agency, 2013.

IAEA Safeguards Glossary, 2001 Edition. Vienna: International Atomic Energy Agency, 2003.

International Conference on Illicit Nuclear Trafficking: Collective Experience and the Way Forward. IAEA-CN-154. International Atomic Energy Agency, November 2007.

International Handbook of Evaluated Criticality Safety Benchmark Experiments. Paris: OECD Nuclear Energy Agency, September 2012.

"Japan AEC Finds Direct Disposal Less Costly than Reprocessing." *Nuclear Fuel* 36, no. 24 (2011): 1.

Johansson, Thomas. "Sweden's Abortive Nuclear Weapons Project." *Bulletin of the Atomic Scientists* 42, no. 3 (1986): 31–34.

Johnson, Shirley. *Safeguards at Reprocessing Plants under a Fissile Material (Cutoff) Treaty.* Research Report 6. Princeton, NJ: International Panel on Fissile Materials, February 2009. www.fissilematerials.org/library/rr06.pdf.

Joint Publications Research Service. *Report: Science & Technology, China.* JPRS-CST-88-002. Springfield, VA, January 15, 1988. www.fissilematerials.org/library/ jprs88.pdf.

Jones, Vincent C. *Manhattan: The Army and the Atomic Bomb.* Washington, DC: Center of Military History, U.S. Army, 1985.

Jungk, Robert. *Brighter Than a Thousand Suns: A Personal History of the Atomic Scientists*. New York: Harcourt Brace, 1958.

Katsuta, Tadahiro, and Tatsujiro Suzuki. *Japan's Spent Fuel and Plutonium Management Challenges*. Princeton, NJ: International Panel on Fissile Materials, September 2006.

Kemp, R. S. "The End of Manhattan: How the Gas Centrifuge Changed the Quest for Nuclear Weapons." *Technology and Culture* 53, no. 3 (July 2012): 272–305.

Kemp, R. S. "Nonproliferation Strategy in the Centrifuge Age." PhD thesis, Princeton University, 2010.

Kemp, R. Scott. "A Performance Estimate for the Detection of Undeclared Nuclear-fuel Reprocessing by Atmospheric 85Kr." *Journal of Environmental Radioactivity* 99, no. 8 (2008): 1341–1348.

Krass, Allan S., Peter Boskma, Boelie Elzen, and Wim A. Smit. *Uranium Enrichment and Nuclear Weapon Proliferation*. London, New York: Taylor & Francis Ltd, 1983.

Krige, John, and Kai-Henrik Barth, eds. *Global Power Knowledge: Science and Technology in International Affairs*. Osiris, vol. 21 (Chicago: University of Chicago Press, 2006).

Lanouette, William. *Genius in the Shadows: A Biography of Leo Szilard: The Man behind the Bomb*. New York: C. Scribner's Sons, 1992.

Larson, E. A. G. *A General Description of the NRX Reactor*. AECL-1377. Chalk River, Ontario: Atomic Energy of Canada Limited, July 1961. www.fissilematerials.org/library/lar61.pdf.

Leventhal, Paul, and Yonah Alexander. *Preventing Nuclear Terrorism*. Lexington, MA: Lexington Books, 1987.

Leventhal, Paul L., Sharon Tanzer, and Steven Dolley. *Nuclear Power and the Spread of Nuclear Weapons: Can We Have One without the Other?* Washington, DC: Brassey's, 2002.

Levite, Ariel. "Never Say Never Again: Nuclear Reversal Revisited." *International Security* 27, no. 3 (Winter 2002): 59–88.

Lewis, John, and Xue Litai. *China Builds the Bomb*. Stanford, CA: Stanford University Press, 1988.

Liberman, Peter. "The Rise and Fall of the South African Bomb." *International Security* 26, no. 2 (Fall 2001): 45–86.

"Limiting the Use of Highly Enriched Uranium in Domestically Licensed Research and Test Reactors," *Federal Register* 51, no. 37. U.S. Nuclear Regulatory Commission, February 25, 1986.

Lindemann, F. A., and F. W. Aston. "The Possibility of Separating Isotopes." *Philosophical Magazine* 37, no. 221 (1919): 523–534.

Lloyd, W. R. *Dose Rate Estimates from Irradiated Light-Water-Reactor Fuel Assemblies in Air*. UCRL-ID-115199. Lawrence Livermore National Laboratory, 1994.

Ma, Chunyan, and Frank von Hippel. "Ending the Production of Highly Enriched Uranium for Naval Reactors." *Nonproliferation Review* 8, no. 1 (2001): 86–101.

Mackby, Jennifer, and Paul Cornish. *U.S.-UK Nuclear Cooperation after 50 Years.* Washington, DC: Center for Strategic and International Studies, 2008.

Malik, John. *The Yields of the Hiroshima and Nagasaki Nuclear Explosions.* LA-8819. Los Alamos, NM: Los Alamos National Laboratory, 1985.

Management of Reprocessed Uranium: Current Status and Future Prospects. IAEA-TECDOC-1529. Vienna: International Atomic Energy Agency, February 2007.

Management of the UK's Plutonium Stocks: A Consultation Response on the Long-Term Management of UK-Owned Separated Civil Plutonium. UK Department of Energy and Climate Change, December 1, 2011.

Manual for Protection and Control of Safeguards and Security Interests. DOE-M-5632.1C-1. Washington, DC: U.S. Department of Energy, Office of Security Affairs, Office of Safeguards and Security, July 15, 1994. www.directives .doe.gov.

Mark, J. Carson. "Explosive Properties of Reactor-Grade Plutonium." *Science & Global Security* 4, no. 1 (1993): 111–128.

Marschak, J. "The Economics of Atomic Power." *Bulletin of the Atomic Scientists of Chicago,* February 15, 1946: 3, 11.

Martinez, J. Michael. "The Carter Administration and the Evolution of American Nuclear Nonproliferation Policy, 1977–1981." *Journal of Policy History* 14, no. 3 (2002): 261–292.

Matos, J. E. "Technical Challenges for Conversion of Civilian Research Reactors in Russia." Paper presented at the NAS-RAS Research Reactor Committee Briefing, National Academy of Sciences, Washington, DC, November 29, 2010.

McFarlane, Harold F. "Is It Time to Consider Global Sharing of Integral Physics Data?" *Journal of Nuclear Science and Technology* 44, no. 3 (2007): 518–521.

McPhee, John. *The Curve of Binding Energy.* New York: Farrar, Straus and Giroux, 1974.

Mian, Zia. *Fissile Materials in South Asia: The Implications of the U.S.-India Nuclear Deal.* Princeton, NJ: International Panel on Fissile Materials, 2006.

Mian, Zia, and A. H. Nayyar. "An Initial Analysis of 85 Kr Production and Dispersion from Reprocessing in India and Pakistan." *Science & Global Security* 10 (2002): 151–179.

Mian, Zia, and A. H. Nayyar. "Playing the Nuclear Game: Pakistan and the Fissile Material Cutoff Treaty." *Arms Control Today,* April 2010, 17–24.

Mian, Zia, A. H. Nayyar, R. Rajaraman, and M. V. Ramana. "Fissile Materials in South Asia and the Implications of the U.S.-India Nuclear Deal." *Science & Global Security* 14 (2006): 117–143.

Mian, Zia, A. H. Nayyar, and M. V. Ramana. "Bringing Prithvi Down to Earth: The Capabilities and Potential Effectiveness of India's Prithvi Missile." *Science & Global Security* 7, no. 3 (1998): 333–360.

Model Protocol Additional to the Agreement(s) between State(s) and the International Atomic Energy Agency for the Application of Safeguards. INFCIRC/540 (Corrected). Vienna: International Atomic Energy Agency, September 2007.

Moore, A. T., ed. *Professional Papers of the Corps of Royal Engineers, Volume XXIX, 1903.* Chatham, UK: Royal Engineers Institute, 1904.

Müller, G., and G. Vasaru. "The Clusius-Dickel Thermal Diffusion Column—50 Years After Its Invention." *Isotopenpraxis Isotopes in Environmental and Health Studies* 24, nos. 11–12 (1988): 455–464.

Muranaka, R. G. "Conversion of Research Reactors to Low-Enrichment Uranium Fuels." *IAEA Bulletin* 25, no. 1 (1984): 18–21.

Musharraf, Pervez. *In the Line of Fire: A Memoir.* New York: Free Press, 2006.

Myers, S. M. "Qualification Alternatives to the Sandia Pulsed Reactor (QASPR) Program Science in Sandia's Physical, Chemical, & Nano Sciences Center (PCNSC): Overview." Albuquerque, NM: Sandia National Laboratories, 2007.

"National Security Directive 61, FY 1991–1996 Nuclear Weapons Stockpile Plan." The White House, Washington, DC, July 2, 1991.

Nayyar, A. H., A. H. Toor, and Zia Mian. "Fissile Material Production in South Asia." *Science & Global Security* 6 (1997): 189–203.

Neff, Thomas L. "A Grand Uranium Bargain." *New York Times*, October 24, 1991, A25.

Nero, Anthony V. *A Guidebook to Nuclear Reactors.* Berkeley: University of California Press, 1979.

Nikitin, Mary Beth. *North Korea's Nuclear Weapons: Technical Issues.* RL34256. Washington, DC: Congressional Research Service, April 3, 2013.

Nolan, Janne. *An Elusive Consensus: Nuclear Weapons and American Security After the Cold War.* Washington, DC: Brookings Institution Press, 1999.

Nonproliferation and Arms Control Assessment of Weapons-usable Fissile Material Storage and Excess Plutonium Disposition Alternatives. DOE/NN-0007. Washington, DC: U.S. Department of Energy, January 1997. www.fissilematerials.org/library/doe97.pdf.

Nuclear Black Markets: Pakistan, A. Q. Khan and the Rise of Proliferation Networks. London: International Institute for Strategic Studies, May 2007.

Obeidi, Mahdi, and Kurt Pitzer. *The Bomb in My Garden: The Secret of Saddam's Nuclear Mastermind.* Hoboken, NJ: Wiley, 2004.

Office of Naval Reactor. *Report on Low Enriched Uranium for Naval Reactor Cores.* Washington, DC: U.S. Department of Energy, January 2014, www.ipfmlibrary.org/doe14.pdf.

Oleynikov, Pavel V. "German Scientists in the Soviet Atomic Project." *Nonproliferation Review* 7, no. 2 (Summer 2000): 1–30.

"On the Goals and the Program of Tests at the Test Site No. 2 in 1953." Council of Ministers of the USSR, Draft Resolution, 1953.

Oshkanov, N. N., O. M. Saraev, M. V. Bakanov, P. P. Govorov, O. A. Potapov, Yu. M. Ashurko, V. M. Poplavskii, B. A. Vasil'ev, Yu. L. Kamanin, and V. N. Ershov. "30 Years of Experience in Operating the BN-600 Sodium-Cooled Fast Reactor." *Atomic Energy* 108, no. 4 (2010): 234–239.

Oxide Fuels, Preferred Option. UK Nuclear Decommissioning Authority, June 2012. www.nda.gov.uk.

Parma, Edward J., Curtis D. Peters, David L. Smith, Ahti J. Suo-Anttila, and Terence J. Heames. *Operational Aspects of an Externally Driven Neutron Multiplier Assembly Concept Using a Z-Pinch 14-MeV Neutron Source (ZEDNA).* Albuquerque, NM: Sandia National Laboratories, September 2007.

Perkovich, George. *India's Nuclear Bomb: The Impact on Global Proliferation.* Berkeley: University of California Press, 1999.

Perry, Alfred, and Alvin Weinberg. "Thermal Breeder Reactors." *Annual Review of Nuclear Science* 22 (1972): 317–354.

Plutonium: Credible Options Analysis. UK Nuclear Decommissioning Authority, 2010.

Plutonium: The First 50 Years. United States Plutonium Production, Acquisition and Utilization from 1944 through 1994. DOE/DP-0137. Washington, DC: U.S. Department of Energy, February 1996. www.fissilematerials.org/library/doe96.pdf.

Plutonium and Aldermaston: A Historical Account. London: UK Ministry of Defence, 2000.

Plutonium Disposition Alternatives Study. Y-AES-G-00001. Savannah River Site, May 2006.

Plutonium Fuel: An Assessment, Report by an Expert Group. Paris: OECD Nuclear Energy Agency, 1989. www.oecd-nea.org/ndd/reports/1989/nea6519 -plutonium-fuel.pdf.

Podvig, Pavel. *Consolidating Fissile Materials in Russia's Nuclear Complex.* IPFM Research Report 7. Princeton, NJ: International Panel on Fissile Materials, May 2009. www.fissilematerials.org/library/rr07.pdf.

Podvig, Pavel. "The Fallacy of the Megatons to Megawatts Program." *Bulletin of the Atomic Scientists,* July 23, 2008. thebulletin.org/fallacy-megatons -megawatts-program.

Polakow-Suransky, Sasha. *The Unspoken Alliance: Israel's Secret Relationship with Apartheid South Africa.* New York: Pantheon Books, 2010.

Polmar, Norman, and Kenneth J. Moore. *Cold War Submarines: The Design and Construction of U.S. and Soviet Submarines.* Washington, DC: Potomac Books, 2004.

Powers, Thomas. *Heisenberg's War: The Secret History of the German Bomb.* New York: Knopf, 1993.

Preliminary Cost Assessment for the Disposition of Weapon-Grade Plutonium Withdrawn from Russia's Nuclear Military Programs. Washington, DC:

Joint U.S.-Russian Working Group on Cost Analysis and Economics in Plutonium Disposition, April 2000. www.bits.de/NRANEU/NonProliferation/docs/usrusswg.pdf.

Press Schwartz, Rebecca. "The Making of the History of the Atomic Bomb: Henry DeWolf Smyth and the Historiography of the Manhattan Project." PhD thesis, Princeton University, 2008.

Primack, Joel, and Frank von Hippel. *Advice and Dissent: Scientists in the Political Arena*. New York: Basic Books, 1974.

Proceedings of the International Conference on the Peaceful Uses of Atomic Energy Held in Geneva, August 8–20, 1955. New York: United Nations, 1956.

Project on Alternative Systems Study (PASS): Final Report. TR 93-04. Stockholm: SKB, October 1992. www.skb.se.

Rahn, Frank J. *A Guide to Nuclear Power Technology: A Resource for Decision Making*. New York: Wiley, 1984.

Ramana, M. V. "An Estimate of India's Uranium Enrichment Capacity." *Science & Global Security* 12, nos. 1–2 (2004): 115–124.

Reed, B. C. "Centrifugation During the Manhattan Project." *Physics in Perspective* 11, no. 4 (2009): 426–441.

Rehman, Shahid-Ur. *Long Road to Chagai*. Islamabad: Printwise Publications, 1999.

Reiss, Mitchell. *Bridled Ambition: Why Countries Constrain Their Nuclear Capabilities*. Baltimore, MD: Johns Hopkins University Press, 1995.

Reistad, Ole, and Styrkaar Hustveit. "HEU Fuel Cycle Inventories and Progress on Global Minimization." *Nonproliferation Review* 15, no. 2 (2008): 265–287.

Report by M.A.U.D. Committee on the Use of Uranium for a Bomb. London: Ministry of Aircraft Production, July 1941. www.fissilematerials.org/library/maud.pdf.

Report to Congress: Disposition of Surplus Defense Plutonium at Savannah River Site. Washington, DC: U.S. National Nuclear Security Administration, Office of Fissile Material Disposition, February 15, 2002.

Restricted Data Declassification Decisions 1946 to the Present (RDD-8). U.S. Department of Energy, January 1, 2002. www.osti.gov/opennet/policy.jsp.

A Review of the Deep Borehole Disposal Concept for Radioactive Waste. N/108. Oxfordshire: Nirex, June 2004.

Rhodes, Richard. *Arsenals of Folly: Nuclear Weapons in the Cold War*. New York: Alfred A. Knopf, 2007.

Rhodes, Richard. *The Making of the Atomic Bomb*. New York: Simon & Schuster, 1995.

Richelson, Jeffrey. *Spying on the Bomb: American Nuclear Intelligence from Nazi Germany to Iran and North Korea*. New York: Norton, 2006.

Richelson, Jeffrey T. *America's Space Sentinels: DSP Satellites and National Security*. Lawrence: University of Kansas Press, 1999.

Rotblat, Joseph. "Nobel Lecture: Remember Your Humanity." Oslo, Norway, December 10, 1995. www.nobelprize.org/nobel_prizes/peace/laureates/1995/rotblat-lecture.html.

Rotblat, Joseph, Jack Steinberger, and Udgaonkar Bhalchandra, eds. *A Nuclear-Weapon-Free World: Desirable? Feasible?* Boulder: Westview Press, 1993.

Roux, A. J. A., W. L. Grant, R. A. Barbour, R. S. Loubsec, and J. J. Wannenburg. "Development and Progress of the South African Enrichment Project." In *Proceedings of the International Conference on Nuclear Power and Its Fuel Cycles; Salzburg, Austria; 2–13 May 1977.* Vienna, Austria: International Atomic Energy Agency, 1977. http://www.iaea.org/inis/collection/NCLCollectionStore/_Public/08/303/8303321.pdf.

Safeguards Statement for 2012. Vienna: International Atomic Energy Agency, 2013. www.iaea.org/safeguards/es/es2012.html.

Sailor, William C, David Bodansky, Chaim Braun, Steve Fetter, and Bob van der Zwaan. "A Nuclear Solution to Climate Change?" *Science* 288 (2000): 1177–1178.

Saunders, Stephen. *IHS Jane's Fighting Ships 2012–2013.* Coulsdon, UK: IHS Jane's, 2012.

Scenarios and Costs in the Disposition of Weapon-grade Plutonium Withdrawn from Russia's Nuclear Military Programs. Washington, DC: Joint U.S.-Russian Working Group on Cost Analysis and Economics in Plutonium Disposition, 2003.

Schaerf, Carlo, Brian Holden Reid, and David Carlton, eds. *New Technologies and the Arms Race.* London: Macmillan, 1989.

Schaff, David P., Won-Young Kim, and Paul G. Richards. "Seismological Constraints on Proposed Low-Yield Nuclear Testing in Particular Regions and Time Periods in the Past." *Science & Global Security* 20 (2012): 155–171.

Scheinman, Laurence. *Atomic Energy Policy in France Under the Fourth Republic.* Princeton, NJ: Princeton University Press, 1965.

Schell, Jonathan. *The Abolition.* New York: Knopf, 1984.

Schneider, Erich A., and William C. Sailor. "Long-term Uranium Supply Estimates." *Nuclear Technology* 162, no. 3 (2008): 379–387.

Schneider, Mycle, and Yves Marignac. *Spent Nuclear Fuel Reprocessing in France.* Princeton, NJ: International Panel on Fissile Materials, April 2008.

Seaborg, Glenn T. "Plutonium: Economy of the Future." *Vital Speeches of the Day* 37, no. 3 (1970): 69.

Sease, J. D., R.T. Primm III, and J. H. Miller. *Conceptual Process for the Manufacture of Low-Enriched Uranium/Molybdenum Fuel for the High Flux Isotope Reactor.* ORNL/TM-2007/39. Oak Ridge, TN: Oak Ridge National Laboratory, 2007.

Serber, Robert. *The Los Alamos Primer: The First Lectures on How to Build an Atomic Bomb.* 1st ed. Berkeley: University of California Press, 1992.

Shaikh, Unis, and M. A. Mubarak. "Radiation Safety around the 'Hot Facilities' at the PINSTECH." *Nucleus* 8, no. 4 (1971): 13–27.

Siddiqi, Rauf. "Khan Boasts Pakistan Mastered Uranium Enrichment by 1982." *Nucleonics Week*, May 20, 1999: 15.

Smith, Alice Kimball. *A Peril and a Hope: The Scientists' Movement in America 1945–47*. Vol. 2. Cambridge, MA: MIT Press, 1970.

Smith, Grant F. *Divert!: NUMEC, Zalman Shapiro and the Diversion of US Weapons Grade Uranium into the Israeli Nuclear Weapons Program.* Washington, DC: Institute for Research, Middle Eastern Policy, 2012.

Smyth, Henry DeWolf. *Atomic Energy for Military Purposes: The Official Report on the Development of the Atomic Bomb under the Auspices of the United States Government, 1940–1945.* Princeton, NJ: Princeton University Press, 1945.

Socolow, Robert, and Alexander Glaser. "Balancing Risks: Nuclear Energy & Climate Change." *Daedalus* 138, no. 4 (2009): 31–44.

Soddy, Frederick. *The Interpretation of Radium.* 3rd ed. New York: G. P. Putnam's Sons, 1912.

Sources and Effects of Ionizing Radiation. New York: United Nations Scientific Committee on the Effects of Atomic Radiation (UNSCEAR), Report to the General Assembly, 2000.

Sources and Effects of Ionizing Radiation. New York: United Nations Scientific Committee on the Effects of Atomic Radiation (UNSCEAR), Report to the General Assembly, 1988.

"The South Asian Nuclear Balance: An Interview With Pakistani Ambassador to the CD Zamir Akram." *Arms Control Today*, December 2011.

Spector, Leonard S. *Nuclear Proliferation Today.* New York: Vintage Books, 1984.

Staats, Elmer B. *Quick and Secret Construction of Plutonium Reprocessing Plants: A Way to Nuclear Weapons Proliferation?* EMD-78-104. Report by the Comptroller General of the United States, October 6, 1978.

Statement on Defence 1956. Cmd 9691. London: HMSO, 1956.

"The Statute of the IAEA." International Atomic Energy Agency, Vienna, 1956. www.iaea.org/About/statute.html.

The Strategic Defence Review. Cm 3999. London: UK Ministry of Defence, July 1998. www.fissilematerials.org/library/mod98.pdf.

Strengthening the Effectiveness and Improving the Efficiency of the Safeguards System and of the Model Additional Protocol. Report by the Director General. GC(54)/11. Vienna: International Atomic Energy Agency, July 27, 2010.

The Structure and Content of Agreements between the Agency and States Required in Connection with the Treaty on the Non-Proliferation of Nuclear Weapons. INFCIRC/153 (Corrected). Vienna: International Atomic Energy Agency, June 1972.

Stumpf, Waldo. "South Africa's Nuclear Weapons Program: From Deterrence to Dismantlement." *Arms Control Today* (December 1995): 3–8.

Summary Volume, International Nuclear Fuel Cycle Evaluation (INFCE). Vienna: International Atomic Energy Agency, 1980.

Takubo, Masafumi, and Frank von Hippel. *Ending Reprocessing in Japan: An Alternative Approach to Managing Japan's Spent Fuel and Separated Plutonium.* Princeton, NJ: International Panel on Fissile Materials, 2013, www .fissilematerials.org/library/rr12.pdf.

Teller, Edward. *The Legacy of Hiroshima.* Garden City, NY: Doubleday, 1962.

Ten-Year Site Plan, 2014–2023. DOE/ID-11474. Idaho National Laboratory, June 2012. www.inl.gov/publications/d/ten-year-site-plan.pdf.

Thorne, Leslie. "IAEA Nuclear Inspections in Iraq." *IAEA Bulletin*, no. 1 (1992): 16–24.

Timbie, James P. "Energy from Bombs: Problems and Solutions in the Implementation of a High-Priority Nonproliferation Project." *Science & Global Security* 12, no. 3 (2004): 165–189.

Treaty on the Non-Proliferation of Nuclear Weapons. INFCIRC/140. Vienna: International Atomic Energy Agency, April 22, 1970.

Trial Calculation of Nuclear Fuel Cycle Cost, Discussion Paper. Japan Atomic Energy Commission, Subcommittee on Nuclear Power and Fuel Cycle Options, October 25, 2011.

Turner, James Edward. *Atoms, Radiation, and Radiation Protection.* New York: Pergamon Press, 1986.

United States Nuclear Tests, July 1945 through September 1992. DOE/NV-209, Revision 15. U.S. Department of Energy, Nevada Operations Office, December 2000.

The United States Plutonium Balance, 1944–2009. Washington, DC: U.S. Department of Energy, June 2012, www.fissilematerials.org/library/doe12.pdf.

Uranium 2009: Resources, Production and Demand. Paris: OECD Nuclear Energy Agency, 2009.

von Baeckmann, Adolf, Garry Dillon, and Demetrius Perricos. "Nuclear Verification in South Africa." *IAEA Bulletin*, no. 1 (1995): 42–48.

von Hippel, Frank. *Managing Spent Fuel in the United States: The Illogic of Reprocessing.* Princeton, NJ: International Panel on Fissile Materials, January 2007.

von Hippel, Frank. *The FMCT and Cuts in Fissile Material Stockpiles.* Geneva: Disarmament Forum, United Nations Institute for Disarmament Research, 1999.

von Hippel, Frank. *The Uncertain Future of Nuclear Energy.* Research Report. Princeton, NJ: International Panel on Fissile Materials, September 2010.

von Hippel, Frank, David H. Albright, and Barbara G. Levi. *Quantities of Fissile Materials in US and Soviet Nuclear Weapons Arsenals.* Princeton, NJ: Princeton University/Center for Energy and Environmental Studies, 1986.

von Hippel, Frank, Rodney Ewing, Richard Garwin, and Allison Macfarlane. "Nuclear Proliferation: Time to Bury Plutonium." *Nature* 485, no. 7397 (2012): 167–168.

Wakeford, Richard. "The Windscale Reactor Accident: 50 Years On." *Journal of Radiological Protection* 27, no. 3 (2007): 211–215.

Walker, J. Samuel, and George T. Mazuzan. *Containing the Atom: Nuclear Regulation in a Changing Environment, 1963–1971*. Berkeley: University of California Press, 1992.

Walker, William. *Nuclear Entrapment: THORP and the Politics of Commitment*. London: Institute for Public Policy Research, 1999.

Wallace, Michael D., Brian L. Crissey, and Linn I. Sennott. "Accidental Nuclear War: A Risk Assessment." *Journal of Peace Studies* 23, no. 1 (1986): 9–27.

Weart, Spencer R., and Gertrud Weiss Szilard. *Leo Szilard: His Version of the Facts: Selected Recollections and Correspondence*. Cambridge, MA: MIT Press, 1978.

Weber, W. J., R. C. Ewing, and W. Lutze. "Performance Assessment of Zircon as a Waste Form for Excess Weapons Plutonium under Deep Borehole Burial Conditions." *MRS Online Proceedings Library* 412 (1995): 25–32.

Wells, H. G. *The World Set Free*. London: Macmillan, 1914. www.gutenberg.org/ebooks/1059.

"White House Press Release on Hiroshima, Statement by the President of the United States." Washington, DC, August 6, 1945.

Wigeland, Roald. *Repository Benefit Analysis*. ANL-AFCI-089. Argonne, IL: Argonne National Laboratory, 2003.

Williams, Robert C., and Philip L. Cantelon, eds. *The American Atom: A Documentary History of Nuclear Policies from the Discovery of Fission to the Present, 1939–1984*. Philadelphia: University of Pennsylvania Press, 1984.

Williams, Robert Chadwell. *Klaus Fuchs, Atom Spy*. Cambridge, MA: Harvard University Press, 1987.

Willrich, Mason, and Theodore B. Taylor. *Nuclear Theft: Risks and Safeguards*. Cambridge, MA: Ballinger, 1974.

Wittner, Lawrence S. *Confronting the Bomb: A Short History of the World Nuclear Disarmament Movement* (Stanford, CA: Stanford University Press, 2009).

Wood, Houston, Alexander Glaser, and R. Scott Kemp. "The Gas Centrifuge and Nuclear Weapons Proliferation." *Physics Today* 61, no. 9 (2008): 40–45.

Wood, T. W., B. D. Reid, J. L. Smoot, and J. L. Fuller. "Establishing Confident Accounting for Russian Weapons Plutonium." *Nonproliferation Review* 9, no. 2 (2002): 126–137.

Wright, Christopher M. "Low-Yield Nuclear Testing by North Korea in May 2010: Assessing the Evidence with Atmospheric Transport Models and Xenon Activity Calculations." *Science & Global Security* 21 (2013): 3–52.

Xoubi, N., and R. T. Primm III. *Modeling of the High Flux Isotope Reactor Cycle 400*. ORNL/TM-2004/251. Oak Ridge, TN: Oak Ridge National Laboratory, August 2005.

Yudin, Yuri. *Multilateralization of the Nuclear Fuel Cycle: Assessing the Existing Proposals*. Geneva: United Nations Institute for Disarmament Research, 2009. www.unidir.ch.

Zarimpas, Nicholas. *Transparency in Nuclear Warheads and Materials: The Political and Technical Dimensions. Stockholm International Peace Research Institute*. Oxford: Oxford University Press, 2003.

Zeng, Shi, ed. *Proceedings of the Ninth International Workshop on Separation Phenomena in Liquids and Gases, 18–21 September 2006*. Beijing: Tsinghua University Press, 2007.

Zippe, Gernot. *The Development of the Short Bowl Ultracentrifuge, University of Virginia, Oak Ridge Operations Report*. ORO-315. Washington, DC: U.S. Atomic Energy Commission, July 1960.

Zoellner, Tom. *Uranium: War, Energy, and the Rock that Shaped the World*. New York: Viking, 2009.

Index

The letter *f* following a page number denotes a figure, *g* denotes a term found in the glossary, and *t* denotes a table.

www.ingramcontent.com/pod-product-compliance
Lightning Source LLC
Chambersburg PA
CBHW031413270326
41929CB00010BA/1433